The New Urban Aesthetic

The New Urban Aesthetic

Digital Experiences of Urban Change

MÓNICA MONTSERRAT DEGEN AND GILLIAN ROSE

BLOOMSBURY VISUAL ARTS
LONDON • NEW YORK • OXFORD • NEW DELHI • SYDNEY

BLOOMSBURY VISUAL ARTS
Bloomsbury Publishing Plc
50 Bedford Square, London, WC1B 3DP, UK
1385 Broadway, New York, NY 10018, USA
29 Earlsfort Terrace, Dublin 2, Ireland

BLOOMSBURY, BLOOMSBURY VISUAL ARTS and the Diana logo are trademarks of Bloomsbury Publishing Plc

First published in Great Britain 2022
Paperback edition published 2023

Cover design by Eleanor Rose
Cover photograph: Waterlicht, Amsterdam, 2017 © Daan Roosegaarde / www.studioroosegaarde.net

A catalogue record for this book is available from the British Library.

Library of Congress Cataloging-in-Publication Data.
Names: Degen, Mónica Montserrat, author. | Rose, Gillian, 1962- author.
Title: The new urban aesthetic: digital experiences of urban change / Mónica Montserrat Degen and Gillian Rose.
Identifiers: LCCN 2021036129 (print) | LCCN 2021036130 (ebook) | ISBN 9781350070837 (hardback) | ISBN 9781350283510 (paperback) | ISBN 9781350070844 (pdf) | ISBN 9781350070851 (epub) | ISBN 9781350070868
Subjects: LCSH: Smart cities. | Technology–Social aspects. | City and town life.
Classification: LCC TD159.4 .D46 2021 (print) | LCC TD159.4 (ebook) | DDC 307.760285–dc23
LC record available at https://lccn.loc.gov/2021036129
LC ebook record available at https://lccn.loc.gov/2021036130

ISBN: HB: 978-1-3500-7083-7
PB: 978-1-3502-8351-0
ePDF: 978-1-3500-7084-4
eBook: 978-1-3500-7085-1

Typeset by Deanta Global Publishing Services, Chennai, India

Contents

Illustrations

Figures

Table

Preface and Acknowledgments

This book is about how digital technologies are changing the sensory experiencing of urban environments. The ways in which cities are planned, designed, represented, and lived are some of the things that have preoccupied us in the last two decades and which we have researched together and separately through various projects. Since around 2010 we noticed from our research a profound change in the working practices and perceptions of those conceiving new urban redevelopments and in the way city users engaged with and discussed their experiences of urban life, as digital technologies were drawn upon more and more often in both the making of and living in cities. In 2017, over a coffee, when discussing our work and academic journeys, we realized that a common theme was emerging from our projects. We identified something new to say about how digital visualizations, from computer-generated images to mobile apps and social media, have transformed both the conception and engagements individuals have with cities. Increasingly everyday sensory urban experience is mediated by digital technologies, and we wanted to explore what consequences this has for how cities are felt and planned. This is how this book came to be written.

While the arguments in this book are new and have not been published elsewhere, our thinking has been informed by a range of research projects we would like to acknowledge, as well as the papers that have been published from those projects. We started working together on our first project between 2007 and 2009, analyzing how people experience two very different town centers in a project called "Urban Aesthetics: A Comparison of Experiences in Milton Keynes and Bedford Town Centres" (funded by the Economic and Social Research Council grant RES-062-23-0223). This inspired the next project on the ways in which computer-generated images transform architectural working practices and the design of future developments, and from 2011 to 2013, with Claire Melhuish, we worked on "Architectural Atmospheres: The Impact of Digital Technologies on Architectural Design Practice" (again funded by the Economic and Social Research Council, grant RES-062-23-3305). A series of papers have been published from these projects that have fed into the thinking and development of this book (Degen et al., 2008, 2017; Degen and Rose, 2012; Melhuish et al., 2014, 2016; Rose et al., 2014, 2015).

Independently, we then continued our research on cities and digital technologies. Gillian developed her interest in the digital mediation of urban spaces by leading the project "Smart Cities in the Making: Learning from Milton Keynes" (ES/N014421/1), also funded by the Economic and Social Research Council. That project looked at a range of different so-called smart projects in MK, and alongside it she wrote a number of more theoretical papers conceptualizing urban space, digital technologies, and posthuman experiencing (Rose, 2016, 2017a, 2017b, 2018, 2019a, 2019b).

Mónica followed her interest in the politics of urban experiences by leading the Arts and Humanities Research Council networking grant between 2015 and 2017 "Sensory Cities: Researching, Representing and Curating Sensory-emotional Landscapes of Urban Environments" (AH/M006379/1) (see also www.sensorycities.com); in 2016 she was awarded a British Academy Mid-Career Fellowship for "Timescapes of Urban Change" (MD140041) (see www.sensescitiescultures.com); and in 2017 she led a research project in collaboration with the Museum of London called "Sensory Explorations of West Smithfield" (see www.sensorysmithfield.com) funded by the Brunel University Research Development Fund. These projects also led to a number of publications (Buckingham et al., 2018; Degen, 2017, 2018; Degen and Barz, 2019; Degen and Lewis, 2020).

These research projects drew on a range of methodologies which produced the data used in this book. Methods of research included walk-along interviews, semi-structured interviews, observational research, photo elicitation, surveys, streetscape analysis, sensory recordings, oral history research, content analysis, and the thematic analysis of photographs, CGIs, and Instagram posts.

In writing this book we are indebted to the support of many colleagues, family, and friends. In particular we would like to thank Clare Melhuish with whom we undertook our project in Qatar and who has coauthored Chapter 3 in this book. Gillian would like to thank past colleagues at the Open University and current colleagues in the School of Geography and the Environment at the University of Oxford; and Mónica, the Sociology and Communication Division at Brunel University. Jennifer Dodsworth helped us make sense of the Instagram data discussed in Chapter 5; and Isobel Ward helped with referencing and formatting. Both Isobel and Jennifer also helped with the creation of the figures in Chapter 5. We also want to thank the researchers who have been involved in our projects over the years, and helped to shape our data collection, our thinking, and our writing: as well as Claire, these include Caitlin DeSilvey, Begum Basdas, Camilla Lewis, Astrid Swenson,

Manuela Barz, Isobel Ward, Parvati Raghuram, Sophie Watson, Nick Bingham, and Matthew Cook.

We would also like to extend our deep thanks to all the people who participated in our various research projects. We interviewed architects and planners, marketing managers and tourists, clubbers and visualizers, and many more. We are very grateful for all their contributions, without which this book would have been impossible.

Gillian would like to thank her daughter Lydia for teaching her about online life and so much more, and Mónica for being an exemplary coauthor, particularly for being able to write excellent conclusions. Stephanie Taylor was an engaged reader of early versions of many papers and chapters—thank you Stephanie. Mónica would like to thank the two men in her life: Chris for passionate debates on the digital, and Daniel for keeping her up to date with the latest developments in social media and making every day such joy. Mónica would also like to thank her friends Sarita Malik, Teresa Waller, Marisa Wilson, Alison Duncan, Emma Wainwright, Jordi Canals, and Carmen Petrus Rivas for their support and discussions over the years. Finally, Mónica would like to dedicate this book to her late father Karl Degen—you always pushed me to go that little bit further, I miss you.

1

Introducing the New Urban Aesthetic

We've planned to meet a friend for lunch in a regenerated area of a large city in Europe. At home, we google the latest restaurant reviews, swipe through Google Map on our phone, look at various restaurants on Instagram, doublecheck hashtags, then book a table using an app. We plan the quickest journey there using Google Maps again. It's a bit too warm to pick up a city bike, so we get a bus, paying our fare with a swipe card, gaze out the window, glance at the bus's animated adverts on digital screens, glance at our phone, again, and emerge into a warm and humid mid-morning. Striding through a newly redesigned streetscape, running a bit late, irritated and slightly disoriented, our eyes frantically scan the phone in our hands, as our finger scrolls this way and that, checking map directions and trying to work out whether we are walking in the right direction, looking out for the restaurant. We pass billboards with computer-generated images showing new buildings in the future streetscape. Surrounding the blue dot on the phone's map there is no sign of the pedestrians that jostle past, the hard pavement underneath our feet, the cars and buses grumbling along the roads. The engulfing labyrinth of real-world buildings is very different from the clean digital lines of the screen. We look up, glance back to the phone and try to relate to the clarity of a blue dot pursuing a gray line with the instruction to turn left and an estimated arrival time to our destination.

Smells of coffee shops and surrounding rubbish, soundscapes of chattering friends, music from car radios and shop fronts, rough textured walls decorated with ripped posters and half cleaned graffiti: the chaotic cacophony of everyday street life mingles with the glowing phone screen and its digital interface.

1 Setting the Scene

Something is changing in cities. Something about how we feel in urban spaces is altering. In 2008, we asked people in a small market town in south-east England to describe its high street, and we posed the same question to visitors to a large shopping mall sixteen miles away. Their responses were richly sensory. They described the sounds, smells, and colors of the different environments, even their textures: one like stroking sandpaper, the other like a tile (Degen and Rose, 2012; Rose et al., 2010). This is what we call an urban aesthetic: a particular sensory experiencing of an urban environment.

In 2018, when one of us asked people visiting and working in an area of London undergoing significant change how they experienced its built environment (Degen and Lewis, 2020), the encounter was somewhat different. A third interlocutor appeared to be present. Around the same time as we were talking to people in the high street and mall, Apple had released the first iPhone in 2007. Ever since, smartphones have become more and more significant in how many cities across the world are experienced and, increasingly, managed. For many city dwellers now, leaving home without a phone (and possibly a spare battery or charger too) is unthinkable. Smartphone applications are part of very many everyday activities in very many cities. Travelling, communicating, eating, socializing, and working all entail embodied interactions with phone interfaces (Barns, 2020; de Souza e Silva and Frith, 2012; Graham et al., 2013). They are "filters, control devices, organizers of social networks, locative technologies, and information access platforms" (de Souza e Silva and Frith, 2012: 4). So, in conversations in that London neighborhood in 2018, people still described full and detailed sensory engagements with the materialities of built space and its activities, just like their counterparts a decade before. But now, their smartphones were intervening in those engagements. Many of those people didn't only talk about the environment, for example: they photographed it. They were snapping their lunch or part of a building to post on Instagram. Others had chosen what to see and do on the basis of social media posts from friends, influencers, and local businesses. And the neighborhood itself was the site of highly crafted digital images, on billboards around sites destined for demolition, picturing the new buildings that would replace them in the near future. Our interviews were interrupted by the beeps and chirrups of phones, and some of those interviewees would have been aware that their phones were also generating data about their activities: their locations, their likes, their networks, and their favorites. Somewhere a database would be collating and analyzing that data, ready to make suggestions of how they could travel home, and of where to eat, shop, and drink on their next visit.

This book is about some of the ways in which digital technologies are part of a shift in the everyday sensory experiencing of urban environments. Those digital interventions are what we refer to as a *new* urban aesthetic. Digital technologies are not particularly new, of course, and cities have always been experienced with the help of technological devices, like maps, for example, or signage (Gordon, 2010; Mattern, 2017; McQuire, 2008). But interacting with a whole range of digital screens—particularly not only phone screens but also consoles, big screens, LED displays, and indeed printed images on billboards that could only have been made using software—is now integral to how many cities are experienced every day. And many of those screens gather data that shapes urban experience, from managing traffic flow to suggesting places to eat. This book explores how the everyday sensory experience of a city is mediated by digital technologies.

For Kember and Zylinska, mediation is "a multiagential force that incorporates humans and machines, technologies and users, in an ongoing process of becoming-with" (2012: 40). This book is about how digital technologies are shaping a particular kind of becoming-with for both people and cities. Mediation "becomes a key trope for understanding and articulating our being in, and becoming with, the technological world, our emergence and ways of interacting with it, as well as the acts and processes of temporarily stabilizing the world into media, agents, relations, and networks" (Kember and Zylinska, 2012: xv). This book discusses a few examples of how some digital technologies and their various users are co-producing particular kinds of urban worlds.

This co-production is caught up in two particular processes shaping cities now, and these processes are integral to the new urban aesthetic that we describe in this book. The first is a whole series of efforts in very many cities in different places, for different reasons and in different ways, to use digital technologies to manage how cities function. This is not particularly new: it can be traced back to the 1950s at least (Halpern, 2015). But ideas about the intelligent city, the smart city, and the platform city have gained increasing traction in the last decade or so. More and more city governments are attempting to gather and analyze big, real-time data to manage infrastructure like traffic, water, and transport. Many corporations have tried to design and sell hardware and software to enable that gathering and analysis, and indeed have proffered their own services to manage it. Social media platforms have provided a means for their users to share information about places in all sorts of ways: posting, reviewing, liking, and sharing (Caliandro and Graham, 2020; Carah and Shaul, 2016; John, 2016; Leaver et al., 2020). Being in the city for many people now is marked by tweets and Instagram posts (Boy and Uitermark, 2017, 2020). Many corporations—particularly social media networks—have been gathering data from that sharing and collating it, using it, and selling it for profit (Zuboff, 2019). Other platforms, from Uber to AirBnB to Deliveroo, broker relations between different users while smaller apps allow users to book taxis, pizzas, and haircuts. In her book *Platform Urbanism*, Barns points to what are now extensive "intermediations" between these many and diverse digital platforms such that "the city itself is rendered as a platform ecosystem" (2020: 121).

The other process affecting many cities now is the effort to sell themselves. Cities are competing on a "global catwalk" (Degen, 2008) for inward investment, for tourists, for spending, for residents. Again, this is a process with a long history. City agencies do this in many ways but over the past twenty years or so there has been a striking emphasis on the atmosphere of a city. Cities increasingly sell themselves on the feelings that they (claim to) generate (Balibrea, 2017; Kavaratzis, 2009). This is a marketing strategy, but it is also shifting the experiencing of a city. Areas are redeveloped in order to generate more income and as that happens their sensory character changes; they are cleaned up, events are organized from pop-up shows to the Olympics in order to generate urban excitement, noise, and buzz. This has been described as "the experience economy" (Pine and Gilmore, 1999) or what this book, following Böhme (2003), will call the "aesthetic economy." Digital tools are now integral to these efforts, and have altered their feel (Degen et al., 2017). From artfully crafted, computer-generated images, to marketing campaigns on Instagram, to tourist selfies and residents' Facebook posts, the feel, the atmosphere, the sensoriality, of cities is being created and shared using digital devices.

This book explores just one aspect of these digital mediations of contemporary urbanism: its sensory experiencing, or what we call its *aesthetics*. We are interested in the "lived and deeply felt everyday sociality of connections, ruptures, emotions, words, politics and sensory energies, some of which can be pinned down to words or structures; others are intense yet ephemeral" (Kuntsman, 2012: 3). This book focuses on the feel of cities, their atmospheres, their sensory experiencing, and the embodiments they invite. As our opening anecdote suggests, much of that everyday experiencing now happens with various digital devices. Barns (2020) also describes the "intimate entanglements" of living and working in a city full of the screens and data flows of digital platforms large and small. This book asks: What does it feel like to inhabit urban spaces mediated by digital devices? How are urban aesthetics mediated digitally? With and for who? What or who is excluded?

We are not alone in posing these questions (Gandy, 2017; Mackrodt, 2019; Sumartojo and Pink, 2019; Thrift, 2014). But we are particularly concerned to think about how these processes are not the same in all cities, or in all parts of the same city, and they are not experienced in the same way by all people. Cities and bodies are mediated differently. Different processes are at work, using different technologies, with different consequences. Those two dynamics—the use of digital technologies in a city and the branding of urban experience—shape specific versions of the new urban aesthetic. The aesthetics of urban life are imagined, designed, and felt in different ways, and different sensoria induce different kinds of bodily experiencing. The main aim of this book is to develop a conceptual framework which is able to unpack different forms of new urban aesthetics in different places.

The next sections examine our three key terms a little more fully, and explain how the book addresses them. The next chapter will explore our theoretical framework in more detail.

2 The Urban: Materialities and Imaginaries

In this book we draw on a particular tradition of urban scholarship interested in the everyday life of cities. There has been considerable interest across recent urban studies in the sensory experiencing of that every day. Sumartojo and Pink, for example, have recently explored what they term urban "atmospheres" as "specific configurations of sensation, temporality, movement, memory, our material and immaterial surroundings and other people . . . how places and events feel and what they mean to the people who participate in them" (2019: 6). In particular they highlight how atmospheres need to be conceptualized as specific to particular places and experienced differently by each individual. Although embodied experience is central to this broad approach, "it is not just bodies that matter on the street, but a vital materiality that runs through and across bodies and things" (Hubbard and Lyon, 2018: 938). Urban atmospheres are constituted precisely through the intermingling of bodies, the environment and technologies.

This book is aligned with this more materialist emphasis in this work, which foregrounds the entanglements between the materiality of the street and the experiencing of its inhabitation (Amin, 2014). In this approach, technologies are important mediators of how urban environments are experienced (McQuire,

2008). Street lighting, for example, allows the city to be visible at night, shifting the times that spaces can be occupied, and the different materialities of light—candles, gaslight, arc lights, LED—give different qualities to what is seen (Ebbensgaard, 2015, 2020; Otter, 2008). Different systems for managing sewage produce different olfactory experiences. A modern shopping center built with glass window walls and marble floors has a different tactility than a high street with a hotchpotch of buildings in brick, concrete, plaster, and wood (Degen and Rose, 2012). And now, as we have emphasized, digital technologies are mediating urban sensoria. Such an approach "render[s] the street as a fluid and fluctuating space" (Hubbard and Lyon, 2018: 939) that is constantly transformed through shifting relations between bodies, technologies, and urban materialities.

In this book we also emphasize the interpretive work of many kinds through which bodies, technologies, and cities come to have particular meanings. Cities are full of human invention and reinvention (Rose, 2017). As the rich and longstanding literature on urban imaginaries insists, the city is "both the actual physical environment and the space we experience in novels, films, poetry, architectural design, political government, and ideology" (Anderson, 2017; Prakash, 2008: 7). Urban branding is one example of an urban imaginary. Our first case study, discussed in Chapter 3, explores an example of urban design and place branding in which digitally created images were central to the vision, the design process, and the marketing. It explores the images created as part of a large redevelopment project in Doha, Qatar. This case study speaks directly to one of the major processes shaping contemporary urban development we have already mentioned: the competition to attract investment. This competition and the marketing it encourages is longstanding and has always involved highly selective images of places, showing them in their best light (Arantes, 2019). Now, though, computer-generated images show images that are somewhat not only realistic—often based directly on architects' plans, as we will show—but also highly glamorous. An entire new industry of digital experience designers is emerging, indeed, who can make photorealistic computer-generated images but also animations, interactive displays, videos, gifs, VR experiences, and indeed everything in-between. Hence we explore the implications of computer-generated images for how urban spaces are visualized.

The two other urban case studies we discuss in this book are both in the UK. One is the city of Milton Keynes (MK), and the other is the London neighborhood of Smithfield. As we will show, in both these cities too, a range of local agencies are involved in efforts at place promotion. Chapter 4 describes how Milton Keynes is pitching itself as a "smart" city, encouraging tech companies to come and trial new technologies like sensor-based traffic management systems, driverless cars, and data hubs, among other things.

Chapter 5 examines how the London district of Smithfield, meanwhile, is being promoted by its local council as a new tourist and cultural destination, the Culture Mile. What these case studies also show, however, is that the work of making sense of digitally mediated cities using digital technologies is not limited to the sorts of professional visualizers explored in our Doha case study. In our discussion of MK and Smithfield we also pay attention to how local residents, workers, and visitors are also actively involved in mediating their experiences of the place using digital devices and the data they generate: in these case studies, smartphone apps.

3 The Aesthetic: Sensations and Staging

Often taking inspiration from the work of Walter Benjamin, many scholars have pointed out that a key feature of the transformation of capitalist economies is the increased role that aesthetization plays in the production and consumption of goods. Central to their conceptualization of aesthetics is a move away from Kantian conceptions of the aesthetic that focus on the cognitive and disinterested engagement with the world to a reclamation of the original Greek meaning of aesthesis: "the perception of the external world by the senses" (Degen, 2008: 38). As these writers suggest, sensory-aesthetic relations are fundamental to the design of very many everyday commodities from smartphones to new apartment buildings.

The expansive sensual logic of capitalism can be traced back to the emergence of consumer culture in the eighteenth century (Howes, 2005; Thrift, 2012). The department store is often discussed as the exemplar of this process:

> With its theatrical lighting, enticing window displays and its floor after floor of entrancing merchandise—"each separate counter . . . a show place of dazzling interest and attraction" (Dreiser, cited in Saisselin, 1984: 35)—the department store presented a fabulous spectacle of consumer plenty and accessibility. (Howes, 2005: 284)

Visual display is central to the department store experience, but other senses are also crucial, particularly touch and smell. Indeed, the "hypersensuality of the marketplace" has become multisensory and more intense in recent decades (Howes, 2005: 290). Since the 1990s there has been a clear trend across various industries to target consumers' senses: from supermarkets creating the illusion of daylight through special lighting, to advertising at bus stops spraying the product's smell onto unaware travelers (Howes, 2005; Lindstrom, 2005; Swedberg, 2011). This has extended from individual commodities to brands. Lash and Lury (2007) identify a shift in the global culture industry from a focus on the characteristics of mass-produced commodities to an emphasis on the experiential dimension of objects as part of brands. They draw particular attention to how brands operate on the level of intensified experiences. Drawing on Benjamin they explain that one "may perceive the painting say, as an object, but what you *experience* is non-objectual—that is colour. This is the experience of an intensity" (Lash and Lury, 2007: 14).

Lash and Lury (2007) extend their analysis to the cityscape too which, they argue, has taken on similar intense qualities (Degen, 2010; Lorentzen, 2009). Here, consumer culture has transformed the urban environment to such extent that urban space becomes space of "multimodal experience, not just that of vision, a space of virtualities and intensities that actualise themselves not as objects but as events" (Lash and Lury, 2007: 15). Place marketing has shifted to place branding. Thus, city councils increasingly draw on a mix of starchitecture, carefully designed urban spaces, and spectacular events to evoke particular experiences and place attachments, supported by a range of diverse digital visuals. Efforts to build a place-based brand "serve to stage, costume and intensify [urban] life" (Böhme, 2003: 72).

This book's three case studies exemplify different kinds of an intensified, multisensory, digitally mediated urban aesthetic. The developers in Doha wanted international investment and recognition. To that end, they employed internationally renowned architects and went to considerable efforts to design not only spectacular set pieces like a huge central square but also the whole development so that it looked like "something one would want on the cover of a magazine," according to the architectural consultant responsible for this task. The designs and its visualization depended heavily on the particular qualities of computer-generated images which could both visualize different design proposals and bathe them in an aestheticized glow of glamour and luxury.

The Smithfield redevelopment project is typical of contemporary place branding. However, the place branding and development agency in charge of the Culture Mile was trying to create a brand not only through marketing materials like computer-generated images but also by intensifying the experiencing of Smithfield as something dramatic. To that end, it planned various events on the ground, from concerts to art installations, with the intention to create site-specific experiences. To be successful these events not only must be experienced in the moment of their happening but they must also appear on Instagram, to be viewed by others who were not there. On Instagram, their images joined many others picturing the Smithfield area in various ways. Our discussion of Smithfield and the Culture Mile place branding demonstrates how the new urban aesthetic is not just a case of "top-down" place marketing—the new urban aesthetic is also part of everyday experiencing of urban spaces.

Our discussion of Milton Keynes looks at an urban aesthetic experience less dominated by the visual (not that the computer-generated images in Doha or London were intended to evoke only visual sensibilities). The chapter looks at a group of smartphone apps designed as part of various smart city projects in the city. It explores how these apps also assume a particular kind of sensory urban experience: mobility. They anticipate a mobile body constantly on the move, and a mobile body that desires efficiency and convenience. All three of our case studies, then, explore somewhat different versions of the new urban aesthetic. They explore different digitally mediated sensory experiences, and in the process, as we will see, entrain certain kinds of embodied experience while marginalizing others.

4 The Digital: Devices and Data

Digital technologies are "inorganically organised objects" that do things with the electronic signals that constitute digital data (Ash, 2015). They are combinations of hardwares and softwares that collect data; clean, integrate, and analyze it; store it and display it; make it mobile; and delete it. Many cities now are saturated with such devices and circulations (Boyer, 1996; Kitchin and Dodge, 2011; McQuire, 2016; Sassen, 2012). Digital sensors in cities monitor air quality, water levels, and traffic flow. Data analytics manage traffic flow and electricity supply and allocate housing (Eubanks, 2017). Bodies are tracked by heat sensors and digital cameras. Cities are mapped by digital satellite imagery. Urban transit systems use smartcards to track journeys and predict capacity. Digital screens—interactive or not, huge or small—offer services, entertainment, and data. Huge amounts of data flow through and between social media platforms, while "disruptive" platforms like Uber or AirBnB offer services while reconfiguring urban transport and residential patterns, and economies (Barns, 2020; Langley and Leyshon, 2017). Meanwhile, local apps proliferate for ordering taxis and pizzas, making hair appointments and ordering repeat prescriptions, reporting potholes, and paying for parking (Rose et al., 2021). Urban imaginaries are shaped by computer-generated images on billboards and YouTube and movies' digital-visual effects. Kitchin and Dodge (2011) coined the term "code/space" to describe spaces entirely reliant on such mediations.

Our attention in this book focuses on just a very few examples of this proliferation of urban digital technologies. We look at computer-generated images, at images on Instagram, and at the data circulations anticipated in a group of smart city smartphone apps. And we are interested in just one aspect of their working: how they are mediating everyday, urban sensoria.

Our starting point for approaching the digital mediation of urban space, therefore, is the sensory experiencing of embodied encounters with screens. Screens are interfaces between bodies and digital data. Smartphones, for example, accompany many people at every moment of their everyday life. Their use is mundane, repetitive, and habitual; they are "functional and sensorial prostheses" for very many bodies (Srnicek, 2014: 83). Data is both delivered and generated through smartphone apps and other digital screens. The smartphone screen, like other screens, is a complex interface between embodied sensations and digital data and software, and mediates the relations between them (Farman, 2015; MacKenzie and Munster, 2019; Rose et al., 2021; Sumartojo et al., 2016).

Approaching urban digital technologies as sensorial allows us to say something about how different kinds of sensual experiences emerge from different kinds of interactions. We highlight the different kinds of skill and labor which different interfaces mediate, for example. In our discussion of the production of computer-generated images for the Doha redevelopment project, we pay attention to what happens at the screens of the visualizers and then to the various screens through which the image files were displayed. In Milton Keynes, we focus on the screens of smartphone apps and again consider their design and also how their designers anticipated they would be used. We also pay attention to the data circulations mediated by smartphone screens and their smart city apps, and explore their consequences for the kinds of embodied users that emerge with those apps. In Smithfield, we pay attention to the smartphone screens in order to explore the work of making Instagram images: what images are taken, filtered, and tagged; displayed and commented on; searched for; reshared; and, perhaps, deleted.

In all our case studies, we are exploring how the circulation of digital data—which might materialize on a phone or a computer, as an image or a video or animation or audio, through apps or other software packages—now mediates the sensory experiencing of cities. All of the digital things we look at invite a particular embodied, sensory engagement. This is the new urban aesthetic. This aesthetic manifests in somewhat different ways. It might be a lushly gorgeous image, or a real-time data flow, or a dramatic Instagram moment. All of these and more are central to the experiencing of many cities now.

5 Differentiation and Power Relations

This book posits the new urban aesthetic as an atmosphere that emerges in many urban spaces. Of course is it not the only thing that shapes urban experiencing. Many other kinds of technologies continue to mediate urban sensibilities (McQuire, 2008). People in cities bring very different knowledges, memories, and feelings to their urban encounters. While the first aim of this book is to draw attention to the intersection of urban space, digital technologies, and sensory experiences, its second aim is to emphasize that these are differentiated. Urban aesthetics condense more or less intensely, in different places, and different ways in relation to different histories, communities, and bodies. It is critical to address those differences, because in them inhere the power dynamics of the new urban aesthetic.

Much of the work on digital urban infrastructure has been very focused on their alignment with power relations; many studies have argued that such infrastructure deepens existing urban inequalities of many kinds. For example, as Graham and Marvin (2001) have forcefully argued in their thesis on splintering urbanism, the digitalization of infrastructures such as communication technologies, water supply systems, transport facilities, to mention just a few, has exacerbated inequalities of access:

> New, highly polarised urban landscapes are emerging where "premium" infrastructure networks—high-speed telecommunications, "smart" highways, global airline networks—selectively connect together the most favoured users and places, both within and between cities. [. . .] At the same time, however, premium and high-capability networked infrastructures often effectively bypass less favoured and intervening places [. . .]. Often such bypassing and disconnection are directly embedded into the design of networks, both in terms of the geographies of the points they do and do not connect, and in terms of the control placed on who or what can flow over the networks. (Graham and Marvin, 2001: 15)

More recent scholarship is focusing on the wide range of organizations which use digital devices of many kinds to harvest data from urban environments, analyzing how not only efficiencies but also profits are sought through the extraction, circulation, transformation, commodification, integration, storage, and reuse of data. This broader context of the "colonization of the [urban] lifeworld through the commodification and extraction of personal information as data" (Thatcher et al., 2016: 992) can be understood as part of what has been called "platform capitalism," in which financial value is extracted from the collection and analysis of data (Finn, 2017; Srnicek, 2016). Attention has focused on "an asymmetric power relationship in which individuals are dispossessed of the data they generate in their day-to-day lives" (Thatcher et al., 2016: 990) with no compensation (Barns, 2020; Graham, 2020; Richardson, 2020a). Discussions of big data in smart cities have thus been joined by critiques of platform urbanism (see, e.g., Barns, 2020; Fields, Bissell, and Macrorie, 2020; Lee et al., 2020; Leszczynski, 2019; Richardson, 2020b; and see the special issues edited by Rodgers and Moore, 2018; Sadowski, 2020). Other work on algorithmic interventions has extended this critique. Many "software systems survey, capture, and process information about people and things in automated, automatic and autonomous ways, making judgements and enacting outcomes algorithmically without human oversight" (Kitchin and Dodge, 2011: x). Indeed, much attention has been given by geographers to the "automatic production of space" by software (Rose, 2017; Thrift and French, 2002), while others have challenged the specifics of database design and the algorithmic processing of spatial and other data (Amoore and Poitukh, 2016; Finn, 2017; Gieseking, 2017; Kitchin, 2017). The algorithmic production of space may define certain neighborhoods as areas requiring surveillance, for example, and others not, and these decisions are often classed and racialized. Finally, many critics have also challenged the technocratic forms of urban governance that such technologies often enact (Cardullo and Kitchin, 2018; León and Rosen, 2020; McNeill, 2015; see for example Vanolo, 2014; Wiig, 2016). Their unaccountable surveillance of individuals and populations has been emphasized (Browne, 2015; Jefferson, 2017; McNeill, 2016; Sadowski and Pasquale, 2015).

A major goal of this book must therefore be to understand how power relations work in the context of the new urban aesthetic. We look at power rather differently from the work on infrastructure and platforms, however. While there is clearly a political economy of the production and circulation of computer-generated images in relation to urban branding and property markets, and the circulation of smartphone data is clearly part of the production and circulation of data more generally in conditions of platform urbanism, we want to emphasize a different kind of power. Instead, this book focuses on how "the process of mediation is also a process of *differentiation*; it is a historically and cultural significant process of the temporal stabilization of mediation into discrete objects and formations" (Kember and Zylinska, 2012: xvi). What emerges from the becoming-with vitality of the new urban aesthetic is always accompanied by "the loss of mediations never to be actualized" (Kember and Zylinska, 2012: 19). Our understanding of the power relations inherent in the new urban aesthetic attends to both what emerges in specific versions of the aesthetic and also what an aesthetic does not allow to become actual and experienced: "[a]n in-cision is also a de-cision" (Kember and Zylinska, 2012: 81).

We elaborate this position by drawing on the work of Henri Lefebvre (1991) and Jacques Rancière (2006). From Lefebvre, we take his well-known triadic notion of space to embed different aspects of spatiality and temporality into the formation of the new urban aesthetic. Embedding this textured notion of space-time allows us to look for friction, contradiction, and other possible experiencing (Jansson, 2013). From Rancière, we take what he calls "the distribution of the sensible" (2006: 7) to focus directly on what is and is not actualizable in a new urban aesthetic. Rancière suggests that power operates through the spatial and temporal organization of sensory experience and that particular spatial-temporal distributions of sensibilities make some things visible, and seen in particular ways, and others not; some things are made audible while others are impossible to hear. Chapter 2 elaborates on our use of Lefebvre, and Chapter 6 develops a critical vocabulary based on Rancière's arguments.

Certain forms of spatiality and temporality have been particularly associated with digital technologies. For example, one aspect of the distinctiveness of digital data very often highlighted is that it is networked (Fast et al., 2018; McQuire, 2016; Munster, 2013). Digital image files, for example, are designed to be distributed (MacKenzie and Munster, 2019; Rubinstein and Sluis, 2008), and this is as much true of the computer-generated images that picture Doha and Smithfield as it is of the Instagram images of Smithfield. The smartphone apps designed as part of MK's smart city activity were also all about sharing data, uploading it, or downloading it via an app. In the discussions of all our case studies, we discuss various networks through which digital data circulates. As we do so, we explore where data moves, and how it encounters different kinds of work done by different bodies, and what happens in that encounter. What kinds of sensory experiences become possible and which do not? What kinds of sensory embodiments are constituted through particular sensorial experiences of networked data circulation?

We also discuss the temporal organization of digital sensibilities. The temporality most associated with digital technologies is the future. Digital technology itself is often imagined as brand new, shaping future possibilities (Kinsley, 2012), and one way cities brand themselves as forward-looking and innovative is to point to their digital smart innovations (see Chapter 4). In his work on how futures are anticipated, Anderson notes that imagining futures requires both imaginative work and embodied experience, because a critical part of envisaging a future is to experience "how a future event might feel" (2010: 786). In the case of urban redevelopment projects, as we have already noted, their branding very much centers on the offer of specific future aesthetic experiences. In the new urban aesthetic, visions of the future are central. Problematizing those anticipated futures, then, is another way of identifying differentiated distributions of the sensible. What kind of future is anticipated? What does that future feel like? Who can feel it that way and whose other futures are rendered impossible?

Networks and futurities are not the only distributions of the sensible that this book will discuss, and they are very much not the only distributions that are relevant to understanding the power dynamics of digitally mediated cities more generally. But such distributions do allow us to exemplify how power works in the new urban aesthetics. They allow us to think about what sorts of bodies and experiences are represented and which are erased, and to ask how such representations confirm or challenge other imaginaries and experiences. And since Rancière does not line up specific configurations of the sensible with particular interests (Dikec, 2015), they also allow us to remain alert to new configurations of the sensible and new forms of digitally mediated urban experience (Rose, 2017).

The book's analytical vocabulary for understanding the power dynamics of specific versions of the new urban aesthetic is elaborated in Chapter 6. There, we explore three of the most significant ways in which the sensory is spatially and temporally organized in our three case studies, and the implications of those distributions for different embodied experiencing. That account is not a comprehensive account of all the power dynamics of the new urban aesthetic. It most likely will not transfer directly to other cities and other forms of digital mediation: but it offers a first step to think about differences. However, given the centrality of the sensory to so much of the digital mediation of cities, we argue strongly that a focus on the distribution of the sensible is a crucial critical tool for understanding the digital mediation of many cities now.

6 Case Studies, Methods, and Evidence

The book builds its arguments from three case studies in three cities. Each case study is drawn from a larger research project and each used a range of different methods to gather its data. The preface to the book details those projects, the colleagues with whom we collaborated as we worked on them, and their funders. Each project interviewed key stakeholders in the various projects we describe. We talked at length to developers, visualizers, council officers, marketing strategists, architects, app designers, master planners, project managers, and others, in their offices and studios and sometimes on building sites. We gathered visual materials too: PDFs of computer-generated images at different stages of development; Instagram posts (we did not do a netnography in the strict sense but followed specific hashtags); still images from promotional websites; and videos from YouTube. We observed the public spaces surrounding Smithfield for several weeks, sat in and observed architectural practices and meetings, videoed the office space of a visualization company for a working day. We stopped people as they strolled around Smithfield and asked them to answer a short survey on the fly. We analyzed all this evidence systematically, looking for recurring themes and topics. And we have published a number of papers from each one, which detail our methods and the evidence for our interpretations. For this book, we have drawn on those in-depth studies in a looser and more interpretive way, as well as on the work of many other urban scholars, in order to make a broader argument about a wider shift in city life now.

7 Conclusion

In conclusion—and to introduce the rest of the book—our notion of the new urban aesthetic explores the digital mediation of sensory urban experience, particularly in the context of place branding and the imagining of future urban life. We are particularly concerned not only to describe and account for this new urban aesthetic but also to elaborate a critical vocabulary that allows us to see the various ways in which particular kinds of social experiences, identities, and relations become possible—or not—as these technologies are deployed.

We do not wish to generalize about the new urban aesthetic from the case studies discussed in this book. Each case study examines a different version of the new urban aesthetic. In the case of the Doha computer-generated images, we describe an aesthetics of glamour, luxury, and sensuality. In MK, the aesthetic of app-generated data circulation is one of flow: of data and of bodies. In Smithfield meanwhile, mediated with Instagram, the new urban aesthetic is about dramatized events and intense locations. Other cities, with different digital mediations, and different imaginaries, will no doubt instantiate other versions of the aesthetic. And these will also require different critical vocabularies to describe their power relations.

With those caveats in mind, we now turn in the next chapter to a fuller exposition of our interpretive framework.

2

The New Urban Aesthetic

A Conceptual Framework

1 Introduction

Much analysis by urban scholars has been devoted to understanding how global economic processes shape and change the physical urban landscape, producing, among other things, a plethora of redevelopment schemes in cities around the globe (Brenner and Theodore, 2003; Harvey, 1989a, 1989b; Sassen, 1991). Less emphasis has been given to the links between this urban redevelopment and the intensification of urban branding strategies (Ashworth and Voogd, 1990; Greenberg, 2000, 2007; Klingmann, 2007), the emphasis in architecture and urban design on choreographing particular sensations in regenerated places (Degen, 2008; Degen and Rose, 2012) and the rise of urban digital technologies (Barns, 2020; De Waal, 2013; Graham and Marvin, 2001; Luque-Ayala and Marvin, 2020). These are uneven processes within cities and across cities globally (Crang et al., 2006, 2007; Datta, 2020). As the previous chapter explained, this book focuses on their intersection and the consequent emergence of what it calls *the new urban aesthetic*. In this chapter we outline our understanding of the new urban aesthetic and develop our theoretical framework.

Inspired by a range of writers (Barns, 2020; Datta, 2020; Lefebvre, 1991; McQuire, 2016; Sumartojo and Pink, 2019; Thrift, 2012), we aim to understand what kinds of new urban sensations and experiences are brought into being through the emergence of various digital technologies in the context of the redevelopment and rebranding of the built environment, in specific times and places. Our starting point is the mediation of the urban environment by digital technologies. As the previous chapter noted, we draw on Kember and Zylinska's (2012) concept of mediation to apprehend and articulate

> our being in, and becoming with, the technical world, our emergence and ways of interacting with it, as well as the acts and processes of temporarily stabilizing the world into media agents, relations and networks. (Kember and Zylinska, 2012: xv)

To paraphrase Fast et al. (2018: 1), in the urban context mediation engenders novel ways of orienting and re-visioning the self and the city, not only in an intimate and embodied sense but also in terms of making and remaking what the city is. As the previous chapter emphasized, we are concerned to explore how digitally mediated urban life is ridden with distinct and complex power relations. We argue that this being-in and becoming-with the technological world necessitates a careful analysis of the particular constellations of spatiality, temporality, and forms of power in urban environments. In this chapter, we outline a conceptual vocabulary that allows us to focus on those aspects of urban aesthetics.

We begin our argument by discussing the intensification of urban redevelopment schemes across the globe, and in particular the creation of new sensations of place as part of place branding and placemaking. The second section of the chapter then turns to how bodily experience is caught up in this staging of urban aesthetics through an array of digital devices. We argue that the orchestration of new sensations in the urban realm has gone hand in hand with the development of digital technologies that make it possible to design, capture, and perceive urban spaces in new ways. The chapter then moves on to the concept of aesthetics, a term used frequently in debates about contemporary urban developments to describe the visual stylization and design of features and landscape (see, e.g., Lindner and Sandoval, 2021). We clarify our use of the term in this book—we use it to refer to differentiations of sensory experiences—and highlight why we need to differentiate between and within various configurations of new urban aesthetics. As the previous chapter noted, considerable attention has been paid in recent years to the notion of urban atmospheres; but little of that work offers an account of how atmospheres enact quite different forms of sensuous experience, or a vocabulary to describe those differences. Drawing loosely on the work of Henri Lefebvre, we suggest an analysis of the urban aesthetic as perceived, conceived, and lived as a way to analyze its distinctive configurations at the intersection of specific urban materialities, imaginaries, bodily sensations, and digital devices.

We then explore the particular form of power enacted by urban aesthetics. As the previous chapter noted, a lot of work on digitally mediated cities has explored the neoliberalism of smart cities and the exploitive political economy of platform urbanism (see Chapter 4). Little of that work has looked at the power relations enacted in digitally mediated urban aesthetics. But, in another moment of differentiation, distinctive forms of urban aesthetics value different sensory experiences differently. Some sensations are prized, others are disliked. As Ngai (2010: 954) notes, articulating an aesthetic is also to make a judgment about pleasure and displeasure, and in the case of the new urban aesthetic, to judge bodies, neighborhoods, streets, and communities. The final section of this chapter therefore turns directly to questions of power, and how power relations rearticulate urban experiencing through the new urban aesthetics. This section draws on the work of Jaques Rancière (2004, 2006) to propose a vocabulary with which to understand how power relations are framed through distinctive spatial and temporal organizations of aesthetic sensation.

2 Branding Cities, Shaping Experiences

In this section we discuss some of the distinctive features of a new urban aesthetic propelled by the radical spatial transformation of many cities in recent decades. The section first discusses the particular characteristics of this new phase of urban restructuring; we then move onto discussing how sensory experiences have become central features in urban branding and placemaking strategies in cities; and end by showing how this is leading to a deliberate choreographing of interactions with and experiences of the city.

Sweeping changes in the nature of capitalism instigated by neoliberal politics in the 1980s have led to an intense period of urban redevelopment in cities (Brenner et al., 2010; Harvey, 2005; Sassen, 2016). Due to the increased financialization of neoliberal capitalism there is an increased push to produce capital through investment in property. This is driven by speculative profit making through private real estate and the property market (Christophers, 2016). The move from localized industrial production to new forms of financialized global investment is accompanied by the transformation from an economy of needs to an economy of desire and dreams which is "increasing activity on the urban level to attract attention, capital, residents and tourists" (Jensen, 2007: 212; see also Raco, 2014). Examples include expanding financial centers such as the City of London, where high-rise buildings such as the Shard or the Cheesegrater compete for attention; the development of cultural quarters such as in Beijing, where military factories have been transformed into the 798 Art Zone; the remodeling of markets once catering for a local population such as the Boqueria Market in Barcelona into gourmet markets for tourists; or the conversion of former dock areas into heritage areas such as in Liverpool. From London, Madrid or Marseille to Rio de Janeiro, Lagos or Shanghai, cities across the globe are refashioning their once-industrial urban landscapes into dazzling experiential designscapes for display on the competitive urban catwalk (Cronin and Hetherington, 2008; Degen, 2008; Zukin, 2011).

While urban renewal has always taken place, it is important to highlight that each urban restructuring period is shaped by its own distinctive aesthetic values and spatial ideologies. What has been particular to much urban change since the 1980s is an explicit focus on stylization and use of urban design to frame not just a visual-spatial identity of a particular neighborhood or development, but more importantly, the overall sensations that create an identity of place and evoke particular embodied feelings of place (Boyer, 1988; Degen and Lewis, 2020; Degen and Rose, 2012). This attentiveness to the experiential impact of neighborhoods and cities needs to be understood as part of a fiercely competitive global economy where city landscapes are under pressure to perform as marketable commodities, as "brandscapes" judged "by their ability to transform the sensation of the subject" (Klingmann, 2007: 6). The emphasis on branding means that cities are now constructing coherent place identities that will appeal through aesthetic taste parameters to particular social groups. Branding consists of constructing particular sets of associations, establishing particular meanings and lifestyles that are linked to the consumption of a product. In the case of the city, this consists of a combination of three key features: first, the tangible building and infrastructure attributes such as for example the attractiveness of an urban landscape or transport and communication networks; second, the non-tangible attributes such as the existence of a vibrant restaurant scene or a "cool" street life (see Kavaratzis, 2009 for an overview); and lastly, as Greenberg (2000: 228) highlights, the shaping of "the images of and discourses of the city as seen, heard, or read in movies, on television, in magazines and other forms of mass [and social] media."

Contemporary urban regeneration and associated branding strategies now go hand in hand with the promotion of neighborhood-based lifestyles (Bonakdar and Audirac, 2019). Hence, the currency that contemporary cities trade in is not just their image but the lived experience that they promise and enact through the design of actual places. The sensuously perceived experience of places is as important as the physical location itself. Ironically, because of the standardization in the architectural appearance of cities (Friedman, 2010; Richards, 2014), intangible features such as local public life or the local "atmosphere" provide the necessary differentiating aspects (and images) for city developers and marketers (see also Balibrea [2017] on how residents in Barcelona have been coopted as important props of the city's brand). While the trend to facilitate or control urban experiences has always been inherent in urban planning and design, what is new is this explicit focus on a conscious engineering in urban design of sensory experiences (Degen and Rose, 2012). Degen's (2008, 2014) work, for example, has shown how during regeneration schemes in Barcelona and Manchester in the 1990s, urban planners, designers, and policy makers combined a similar series of design, use, and architectural strategies in both cities to orchestrate particular sensations and atmospheres in areas undergoing urban regeneration and aimed at a white middle-class taste (Summers, 2021). Some critics argue that this process has further intensified since 2010, as urban design has gradually morphed into "placemaking," the "deliberate and purposeful approach to place creation" (Lew, 2017: 456), or to put it bluntly, the strategic shaping of the physical, social, and sensory character of neighborhoods (Balibrea, 2017; Markusen and Gadwa, 2010; Nicodemus, 2013). The work of Lew (2017), for example, illustrates how the Singapore Tourism Board has recently listed placemaking as one of its principal activities and "works with public and private stakeholders in several neighbourhood precincts to coordinate infrastructure improvements, events and marketing campaigns 'to improve visitor experience and inject vibrancy to bring the precinct to life' (STB, 2016:1)" (Lew, 2017: 454). However, it is important to appreciate that this is not always a straightforward process: such strategies can be ignored or be resisted.

The importance of enhancing experiences in this aesthetic economy links to Lash and Lury's (2007) identification of a clear shift in the global culture industry from a focus on what they term the extensity of manufactured mass production, to intensity, which is their term for the emphasis on the experiential aspects of objects when turned into brands. They draw particular attention to how brands operate on the level of intensified experiences. Drawing on Benjamin they explain that one

> may perceive the painting say, as an object, but what you *experience* is non-objectual—that is colour. This is the experience of an intensity. Brands may embrace a number of extensities but they themselves are intensities. (Lash and Lury, 2007: 14)

To put it simply, while brands might become distributed globally as products, it is the intensity of the experience they evoke that entices their consumption.

Important for our argument is that Lash and Lury extend their analysis of branding to the cityscape which has taken on similar intensive qualities. Here, consumer culture has transformed the urban environment to such an extent that urban space becomes space of "multimodal experience, not just that of vision, a space of virtualities and intensities that actualise themselves not as objects but as events" (Lash and Lury, 2007: 15). Crucial here are the memorable moments which city branding campaigns must create and evoke in order to sell themselves to potential investors and tourists by providing not only an enticing physical environment but also an array of experiences through atmospheric places, activities, festivals, or events. What is desired is a vibrant urban environment: one which can be captured (digitally) and then distributed globally across traditional, and, increasingly, on social media. Thrift suggests that such an emphasis on the experiential is linked to capitalism's constant need for reinvention. This new economic system is reliant on the expansion of the technologies and practices of an "expressive infrastructure" that will "act as a pipeline for affect and imagination" between consumers and products and between urban inhabitants and the city (Thrift, 2012: 144).

Like Thrift, we are interested in the roles that digital technologies play in this expressive infrastructure in which consumers' emotions and perception are intimately and responsively tracked to create a "rolling world" of sensations (Thrift, 2012: 142), and which forms the larger context for the new urban aesthetic. We explore three examples of this digital mediation of urban sensation through the book's three case studies. But all our case studies discuss perhaps one of most obvious ways in which software and hardware is deployed to produce an expressive urban world: computer-generated images of urban redevelopment projects. More-or-less photorealistic computer-generated images (CGIs) are now the most common type of image used to visualize and market future urban redevelopments and their envisaged social uses (Degen et al., 2017). The book's most extended discussion of their use is in Chapter 3, which explores how CGIs were used in the design and marketing of a large urban redevelopment project in Doha. The central purpose of those CGIs was to intensify the experiential aspects of a redevelopment scheme. While drawings of planned buildings always attempted to make them look good, CGIs can produce particularly intense, highly atmospheric, beautiful views of buildings, glowing spotlessly at dusk (Degen et al., 2017). They have become such a ubiquitous part of producing and marketing urban landscapes in contemporary consumer culture that we could claim that they are one way in which cities "are beckoned into existence by code" (Thrift and French, 2002: 311). Hence we give them sustained attention across this book.

This section has highlighted the digital mediation of city branding, urban redevelopment, and sensory experience in the making of the new urban aesthetic. We argue that the branding of contemporary urban redevelopments strongly emphasizes and promotes the sensory appeal of such places and thus attempts to shape people's uses of and attachments to place. In particular, we have suggested how this process has deepened with the development of new digital technologies which shape how places are designed across the globe. Both processes co-constitute specific urban sensoria.

3 Urban Experiencing and Digital Mediations

Senses are central ingredients in the organization of quotidian experience. The senses engage us with the city in that the interplay of different sensory perceptions in places contributes to our spatial orientation, frames our awareness of spatial relationships, and facilitates the appreciation of the qualities of particular places (Degen, 2014). These fine-grained sensations build up into an urban lived experience where the physicality of the city constantly interacts, supports, and collides with our bodies. It is these sorts of sensory experiences that recent urban branding strategies attempt to evoke. In many urban spaces, digital technologies also engage the human sensorium. Many of those technologies have visual interfaces, for example. Advertising hoardings show CGIs. Krajina (2014) has mapped the ways people in city streets look at—and ignore—digital screens both large and small. But the digital mediation of human embodiment is not only visual. In fact, it is multisensory. Screens also often stream sound too, for example. Smartphones remain phones and emit call tones and voices, as well as the pings and buzzes of notifications, and they stream music and soundtracks, so we hear digital technologies on city streets as well as look at them. Touchscreens offer smooth surfaces to our fingers as we scroll, tap, and zoom. Portable digital screens have invented new hand gestures: the pinch out, the pinch in, and the swipe (Berry et al., 2013; Krajina, 2014; McQuire, 2010). Many bodies now navigate city streets staring at smartphone screens while messaging or using apps like Google Map and City Mapper (Verhoeff, 2012; Wilmott, 2016), the phone converting them into a location as they do so (Fast et al., 2018). It's even been suggested that this is producing a new form of urban body entirely: the "smartphone zombie" (The Independent, 2017).

What we are describing is the way in which digital devices are increasingly part of the sensory texture of everyday urban life. Bodies hold, carry, and touch smartphones, tablets, laptops, gaming devices, ebook readers, smartwatches, and fitness trackers, and bodily dispositions and feelings in urban spaces emerge in relation to these devices as much as they do to the buildings, weather, crowds, and signage. What bodies feel in urban spaces is done in relation with these devices as well as with the built environment. Moreover, we can say that bodies in many city spaces now simultaneously occupy both material urban spaces and digital, onscreen environments, which offer more sensory experience. This is what has been called the "'doubly digital' quality of contemporary media" (Moores, 2014: 204). There is an intimate relation between the experience of being in online environments and the corporeality which enacts that experience. A body tapping or swiping on a smartphone is both in a street, say, and it is also in the virtual space of the screen. This doubling is not a merging or synthesis but what Munster calls a "a kind of graft, which is an unequivocal mark of connection and difference" between the fleshy body and virtual body (2006: 23). "New media entice bodies to venture towards incorporeal flows of information and combine, in convergent and divergent ways, the capacities and functions of carbon materialities with those of information flows" (2006: 23). This differential combining is what Munster (2006: 25) terms "digital embodiment."

This doubled digital embodiment means that bodily sensations are always-already entrained in the mediations of urban digital technologies. Smartphone screens are not only read, they are felt: their texture, their glow, their weight. Different apps engage different bodily sensations: mobility in the case of CityMapper, for example, or heart rate and speed in the case of a running app. As bodies are converted into data, as bodies are felt through and with data and devices, the data generated shifts sensorial experience. Thrift (2012: 141) emphasizes the always-emergent feel of this "continually involving and evolving" urban experience. Efforts to brand cities using digital media are part of this wider digitally mediated urban sensorium. Computer-generated images of urban building sites project a distinctive seamless sheen from billboards, for example (Degen et al., 2017); cyclists interact with the data they generate with apps like Strava (Sumartojo et al., 2016); traffic flow is regulated by algorithms. This is a key aspect of how mediation is often suggested to be an environment (McLuhan, 1964; McQuire, 2008). If "neither home nor street nor city can now be thought apart from the media apparatus" (McQuire, 2008: 7), that is now in part because the sensorialization of both city and body wrought with digital mediation.

Some of the consequences of this relationship between digital mediation and bodily sensations are raised in a study of how different neighborhoods in Amsterdam appear on Instagram (Boy and Uitermark, 2017). Boy and Uitermark (2017) argue that the photographs shared on Instagram must be understood as part of the wider changes in Amsterdam's urban spaces, and they point in particular to the way that what and how neighborhoods are shown on Instagram is part of the gentrification of the city. Particular neighborhoods appear on Instagram as desirable in particular ways. Each neighborhood on Instagram appears a little differently: cool music there, hip food there, woke art there. They emphasize that these images are highly aestheticized. As well as deploying the particular visual styles of Instagram, they are attractive images of "refined beauty and good vibes," of delicious food and stimulating art and chic clubs posted by "happy, healthy and hip" people (Boy and Uitermark, 2017: 617). The images imply corporeal sensations that are pleasurable and seductive. They are an expressive infrastructure of "an interminable stream of peak moments" (Boy and Uitermark, 2017: 617).

All three of the case studies we discuss here explore digitally mediated sensoria of different kinds, but our second case study addresses the digital mediation of embodied experience in some detail, using the example of the smartphone applications that have been designed as part of the efforts of Milton Keynes's to go "smart." This discussion explores a specific example of how "the sensory and contingent plane of living bodies is doubled and variably reconfigured through computational schemas" (Munster, 2006: 81). Many smartphones carry apps that mediate bodily experience: fitness trackers, food delivery apps, and photo-sharing apps. As we will explore, despite their ostensibly quite various purposes, Milton Keynes's smart city apps mediate bodies in a quite particular way.

Bodies in cities, then, inhabit a space grafted between the experience of being in online environments and the corporeality which enacts that experience. Bodies do things with digital devices and the data that this doing generates then co-constitutes those corporeal bodies, as for example a body thrills to a particularly beguiling Instagram post. And this shifts how cities are experienced. They are encountered sensorially. This understanding of the sensory mediation of urban space shifts attention from individual embodiments and toward the extension of sensory feelings in and through digitally mediated urban environments: and this is precisely what we are calling the new urban aesthetic.

4 Aesthetics: Produced, Conceived, Perceived, Lived

The original Greek meaning of *aisthesis* is "the perception of the external world by the senses" (Degen, 2008: 38). We have already noted that a range of scholars have highlighted how capitalism in late modernity has increasingly profited from the sensory experiencing of both commodities and of city spaces. Indeed, there have been several extended accounts of urban sensation in that context. These have tended to conceptualize that experiencing in terms of affect (e.g., Ernwein and Matthey, 2018; Latham and McCormack, 2004) or atmosphere (e.g., Bissell, 2010; Buser, 2017; Degen and Lewis, 2020; Gandy, 2017; Sumartojo and Pink, 2019). In this section, we explain how we build on that scholarship by developing a particular notion of aesthetics.

Like affect and atmosphere, aesthetics refers to the sensory. However, the sensory is not simply what individual human bodies feel. The sensory is the relation between the embodied and the material environments. It is a spatially extensive relation. It is what the CGIs mentioned in the first section of this chapter try to visualize: not just what a group of new buildings will look like but the feel that emerges between it and its surroundings and its inhabitants. Recent discussions of urban atmospheres also understand atmospheres as relational sensations between bodies and material places (Anderson, 2009; Degen et al., 2017; Edensor, 2012; Gandy, 2017; Simpson, 2016; Sumartojo and Pink, 2019).

This extensive quality of urban sensations is central to the conceptualization of atmosphere by Gernot Böhme (2003). Böhme's starting point is closely aligned to the account of urban redevelopment given in a previous section. He is interested in atmospheres because he sees atmospheres as central to the capitalist production of sensory pleasures. He describes contemporary capitalism as an "aesthetic economy" which relies on the production of aesthetic value for its profits (Böhme, 2003), and suggests it has led to

> a new resulting aesthetics [that] is concerned with the relation between environmental qualities and human states. This "and," this in-between, by means of which environmental qualities and states are related, is atmosphere. (Böhme, 1993: 114)

Böhme (1993) elaborates that atmospheres should not be conceived as free-floating energies but as sensory attributes that proceed from and co-constitute things, persons, and their relations. Urban experience is constituted through an array of embodied, sensory encounters with things that extend and radiate toward us (Böhme, 2013, 2017). Atmospheres are therefore spatial: they extend and envelop (Anderson, 2009; Frers, 2016; Thrift, 2012). More recent academic discussions of urban atmospheres have focused on their dynamic and often elusive qualities, regarding them as emergent and ephemeral: "atmospheres are perpetually forming and deforming, appearing and disappearing, as bodies enter into relation with one another. They are never finished, static or at rest" (Anderson, 2009: 79). But they do nonetheless have a kind of shape, channeled in part by the expressive infrastructures of digital practices and technologies.

Böhme (2017) is particularly interested in what he calls the "staging" of atmospheres. He is clear that the design of an object—whether a building or a handbag—is not sufficient to project an atmosphere. Rather, objects designed to induce atmospheres have to be positioned, arrayed, sensuously choreographed, and displayed in particular arrangements. Thus, contemporary atmospheres offer "scenes, life spaces, charisma [. . .] what matters are their radiance, their impressions, the suggestions of motion" (Böhme, 2014: 45). This notion of staging is very useful to understand urban branding, as Chapters 3 and 5 will elaborate. It is also relevant to the study of Amsterdam on Instagram referred to earlier in this chapter, which discusses how specific locations—a café, a club—are pictured as scenarios with what Böhme would describe as props, such as a cup of latte art or a striking light effect: "Instagram users selectively and creatively reassemble the city as they mobilise specific places in the city as stages or props in their posts" (Boy and Uitermark, 2017: 613). The emphasis on the arrangement and sensuous display of things focuses attention on the way that an atmosphere is generated as a relation between multiple components. These can be all sorts of things, not only buildings, benches, trees, people, events, light, and so on, but also personal memories (Degen and Rose, 2012), memories of films, social media posts, or advertisements, and what appears on the many screens that now punctuate urban spaces (Berry et al., 2013; Krajina, 2014; Mcquire, 2010; Papastergiadis et al., 2013). Many different elements come together and contribute to the creation of an urban atmosphere.

Böhme is also interested in the production of atmospheres. He repeatedly emphasizes what he calls the "aesthetic labour" that goes into the making of aestheticized commodities with the aim of creating atmospheres when they are put to use. When exploring the new urban aesthetic, it is important to remember that it is labored over. Staging an urban aesthetic requires different kinds of work, by different people, in different places. CGIs, for example, are created by an elaborate division of labor between different sorts of design professionals in different places, as we discuss in Chapter 3 (Rose et al., 2014; and see Chung, 2018). Creating digital visualizations of a planned redevelopment project, or designing an open-access urban data hub, or a smartphone app that uses geolocated data to offer urban services, are themselves social practices which emerge through "localized contexts of practical action" (Cockayne, 2016; Georgiou, 2013; Rodgers et al., 2014: 1063). As well as technical skills, these actions involve sensorial judgments and reactions. A CGI oozing glamour may evoke certain feelings on a billboard—for example envy, desire, or derision—but it may have other effects as it is being labored over in a visualization studio including boredom, cynicism, and exhaustion.

Aesthetic labor, then, is one example of the notion of doubled digital embodiment discussed in the previous section. It considers what kinds of corporeal sensoria are enacted with digital devices as things are made and done with them. This is a particularly useful emphasis because it grounds digital media in a range of different urban institutions, conventions, and practices, and this allows a recognition that their effects may be multiple. So our case studies here look not only at professional designers of different kinds creating—or attempting to create—new urban aesthetics, but also at the ways in which many ordinary users of smartphones also participate in aesthetic production in their everyday practices. Indeed, as Rodgers, Barnett, and Cochrane (2014) argue, it is important to move away from an account of (digital) media as something extrinsic to urban life, and instead to position them as intimately entangled within many kinds of urban social practice (see also Zukin, 2020 on urban tech economies).

This discussion of aesthetic labor introduces a critical aspect of our formulation of the new urban aesthetic. There are different configurations of the new urban aesthetic, which can induce different effects; and their sensuousness may entrain different bodies differently. The work of a visualizer laboring to glamourize a new apartment block is not the same as someone tweeting a smartphone snap in protest at that same development (see also Butler et al., 2018). This brings us to a criticism that has been made against the notion of atmosphere specifically. Gandy, for example, points to tensions between descriptions of "precise characteristics of urban atmospheres associated with specific times, spaces or situations . . . and an implicit attachment to a bounded, idealised, and to some degree ahistorical subject" (2017: 354). Indeed, a common theme in much of the literature on "urban atmospheres" is that "the felt body is seemingly devoid of gender or any other kind of social difference, or indeed any clear sense of historical and geographical context beyond the confines of the (late) modern European city" (Gandy, 2017: 369) (and this is certainly a criticism that could be made of much of Böhme's work). Similar criticisms have also been made of discussions of space that rely on notions of "affect" (see, e.g., Tolia-Kelly, 2006).

This critique is our main reason for using the term "aesthetic" rather than "atmosphere." One of the main aims of this book is to offer a concept of urban sensoria that embeds multiplicity and differentiation. The new urban aesthetic is diverse. Each of our case studies identifies a different configuration of the new urban aesthetic in the three places they discuss. But also, there are aestheticized differences within cities. As the example from Amsterdam shows, different parts of the city are sensorialized differently; and as Böhme's emphasis on aesthetic labor suggests, particular configurations of the new urban aesthetic will enroll some sensory experiences, but may also be felt as alien to some.

In fact, this book makes four moves to embed differentiation into its central concept of the new urban aesthetic. The first is to insist that the new urban aesthetic is configured differently in different places. The second is to be attentive to the work done by corporeal bodies to mediate aesthetically urban spaces: that is, to pay attention to different kinds of aesthetic labor and to the different experiences of digitally mediated urban space that result. The third is to work with Jacques Rancière's notion of the "distribution of the sensible," and we turn to that in the next section of this chapter. The fourth is to suggest that aestheticized urban environments are inherently differentiated: they are *textured*, to use the term deployed by Jansson (2013). To develop his argument, Jansson (2013) turns to the work of Henri Lefebvre (1991). We also draw loosely on Lefebvre's work and specifically on his account of the production of space.

We have already argued that the new urban aesthetic is spatial. Not only does it mediate urban materialities, it does so by creating a relational space between things: the new urban aesthetic is an atmosphere, a stage, an envelope, a radiation, an environment. Following Lefebvre, we should understand these spaces as triadic. They have three components: they are *conceived, perceived*, and *lived*. While Lefebvre elaborates his argument partly in Marxist terms, we draw more on his interest in differentiated phenomenological experience as a central feature of the shaping of the new urban aesthetic (Kincaid, 2020). Indeed, Lefebvre views space as an ongoing practice constructed through bodies engaged in the building, making, representing, framing, and experiencing of space (Simonsen, 2005; Simonsen and Koefoed, 2020). We also suggest therefore that the new urban aesthetics can be described as conceived, perceived, and lived, in order to better describe how it might be differently experienced (see Degen, 2018).

The first element of the spatiality of mediation is *perceived* space. Perceived urban space refers to directly experienced surroundings: the concrete physical and social materiality of a street, or a technological device, or a body, as they are encountered through time. This is the focus of Chapter 4's discussion of smartphone apps. *Perceived urban aesthetics* refers to the everyday relation between bodies, urban built environments, and "the more material, sensuous dimensions of the media; the very stuff in terms of tools and infrastructures for mediation that make up our everyday environments" (Jansson, 2013: 282). Lefebvre defined the perceived as the observable patterns of everyday life—perceived space is "directly sensible and open, within limits, to accurate measurement and description" (Soja, 1996: 66)—which "embodies the interrelations between institutional practices and daily experiences and routines" (Simonsen, 2005: 6). To return to Boy and Uitermark's (2017) discussion of Instagrammed Amsterdam, perceived urban space shows the physical streets and clubs, the bodies inhabiting those locations and making the posts, and the Instagram app on their phone. These are (some of) the perceived materialities of that particular new urban aesthetic.

In Lefebvre's account, *conceived* urban space refers to the rationally abstracted forms of urban spatiality which are mentally conceived in verbal, visual, or written representations. Conceived space is constituted when "knowledge of [urban] material reality is comprehended essentially through thought, as *res cogito*, literally 'thought things'" (Soja, 1996: 79). Jansson suggests that this aspect of urban space (though he uses the term "mediatization") takes the form of the circulation of representations of spaces and places (Jansson, 2013: 283). Such representations "actively transform the world we live in" (Soja, 1996: 67). In many accounts, including Lefebvre's, conceived urban space is created by the "dominant discourses" produced by city planners, developers, and governments and is articulated through things like plans, policies, maps, and photographs (Simonsen, 2005). As we have noted, city plans and developers' proposals also now often include a range of digital images showing not only maps but also photorealistic images of the future planned environment. Böhme's (2014, 2017) account of aesthetic labor focuses on the professional production of such representations. Chapter 3's discussion of the production of CGIs emphasizes the conceived quality of their new urban aesthetic.

But the conceived should also refer to everyday conceptual imaginaries that individuals and the general public might have of places. This is particularly necessary in an era of social media. Social media platforms, whatever else they may do, also provide tools for their users to interpret, represent, and share their interpretations of their urban experiences, as we discuss in Chapter 5. The conceived elements of the new urban aesthetic are produced by many different kinds of urban actors (see, e.g., Boy and Uitermark, 2017; Datta and Thomas, 2021; Graham et al., 2013), and as we will demonstrate, they can complement or contradict each other. We therefore extend Böhme's notion of aesthetic labor and use it to refer to this ongoing interpretive work by all of a city's inhabitants, not just the design professionals working there. As Soja highlights, this conceived "imagined geography tends to become the 'real' geography, with the image or representation coming to define and order the reality" (1996: 79).

What this aspect of space does is as important as who produces it, however. Conceived urban space gives particular values to different elements of the urban realm as a physical and social space; it sorts and classifies areas and bodies into categories. Here we might recall the Instagram images of Amsterdam, which not only select particular parts of the city to post about but also evaluate those neighborhoods positively, as hip and cool. In our discussion of the new urban aesthetic, then, we do not want to lose sight of the importance of the shaping of sensory experience by cultural meanings and values (see also Anderson, 2009; Sumartojo et al., 2019; Sumartojo and Pink, 2019). City spaces are experienced in part because of cultural frames of meaning and processes of interpretation and classification. These are "the interpretive grids through which we think about, experience, evaluate, and decide to act" (Soja, 2000: 324). City spaces are interpreted as well as felt. The conceived aspects of urban spaces have particular qualities emphasized and others (de)valued. The next chapter's discussion of Doha emphasizes the conceived aspects of its new urban aesthetic.

The third category of urban aesthetics, *lived* urban space, is again drawn loosely from Lefebvre. In Lefebvre's work, lived urban space refers to consistent and habitual enactions of urban life. Jansson (2013) describes the lived aspect of mediation as what saturates everyday life and holds it together as a normal way of doing things. Lived space is produced by social practices, which bring together both interpretive grids with bodily experience. Böhme (2014: 48) notes that "the atmosphere of a city is precisely the way that life goes on within it." Thinking again about Instagram, the lived aspect of the mediation of a city like Amsterdam refers to the normalization of using Instagram itself as part of everyday life. In this book, Chapter 5 emphasizes the lived aspects of the new urban aesthetic generated around Smithfield Market. These enactments generate both sensations and interpretations which can challenge the conceptions of other urban actors.

Lefebvre was clear that the perceived, lived and conceived are not distinct and different forms of urban space. They influence each other. Particular combinations of these three aspects of a space create that space's texture, to use Jansson's term:

> A texture is like a fabric with a particular pattern and a particular feel—and thus something more than a text or a sign system. Textures support our sense of continuity and belonging, not only at the representational level, but also in a deeply embodied sense as we learn how to move and act in various settings. (Jansson, 2013: 286)

Each of the three spatial elements of the new urban aesthetic can differ in different places, and they can combine in different ways in different places thus producing different sensorial experiences. These combinatory textures thus give us a way to describe differentiations of and in the new urban aesthetic.

Moreover, the three aspects of this spatial triad can align, but conceived, lived, and perceived spaces can also contradict each other in lived urban space. This triadic conceptualization thus gives us another way to embed differentiation into our account of the new urban aesthetic:

> If we want to study the mediatization of a certain domain of social life, or certain sociospatial arrangements, we must aim for a holistic view that captures the composite effect that occurs through the interaction between perceived, conceived and lived spaces. At the same time, conceptualizing mediatization in this way makes it possible to discern how developments within one spatial realm may successively reinforce and spark off alterations within another realm. (Jansson, 2013: 283)

This composite effect may cohere. Or it may be contradictory, as when planners and campaigners hold different versions of a place's atmosphere (Mackrodt, 2019). This aspect of the new urban aesthetic is explored in Chapter 6.

This section has explained our use of the term "aesthetic" rather than atmosphere in our discussion of how urban redevelopment projects are most often mediated now. Without doubt, that mediation—very often done using digital visualization tools—is part of the capitalist aesthetic economy. We follow Böhme's (2014, 2017) emphasis on the importance of staging to creating atmosphere. Given the imbrication of the new urban aesthetic with the particular imperatives of urban branding, however, we have elaborated his position in order to conceptualize the new urban aesthetic as differentiated. We concur with his emphasis on the aesthetic labor that is done to create the objects with which specific urban aesthetics are staged but add an emphasis on the diverse forms of that labor. But we develop his work by insisting on the differentiation of atmospheres which is generated by the three intersecting qualities of their spatial extension: conceived, perceived, and lived. Atmospheres are therefore complexly textured. They can be multiple and contested. Conceived, perceived, and lived spaces can converge but also diverge: they can be produced with different effects. The next section turns to our final tactic for embedding differentiations in the new urban aesthetic.

5 The New Urban Aesthetic, Difference, and Power

So far, the chapter has outlined the new sensorial dynamics that are emerging through the interrelationship between diverse digital technologies which mediate urban materialities, the redevelopment of urban landscapes, and the sensory experiencing of city spaces. We have established that we conceptualize this new urban aesthetic as always *differentiated*. So far we have discussed this differentiation by suggesting that the new urban aesthetic feels different in different places, that is made by different kinds of *aesthetic labor*, and that its spaces are *textured* because *triadic*. They are perceived in manifold ways through diverse bodies, urban materialities, and technologies; conceived in multiple ways by different urban actors; and lived through manifold practices both everyday and specialized. In this section, we pay direct attention to the power relations articulated through a new urban aesthetic by thinking through notion of the *distribution of the sensible*. This is the fourth move we make to theorize the new urban aesthetic as differentiated.

As argued at various points of this chapter, aesthetics, understood as sensory experiences, have become important components of the production, branding, and experiencing of cities since the late twentieth century. Many urban scholars have understood this shift as simply the latest ideological cloak hiding the harsh reality of ongoing urban exploitation. The branding of new urban redevelopment projects is seen as the

> production of immaterial values of economic significance . . . [where] not only . . . architecture serve brands, but architects themselves have become brands. . . . Function in architecture also expands until it reaches progressively less utilitarian sensory dimensions. Architecture must provide unique and memorable experiences as part of its business model. For this a wide range of materials, building skins, chromatic effects, lighting systems and digital ornaments are mobilised. (Arantes, 2019: 18–19)

As the previous chapter outlined, smart cities have been criticized for exemplifying technocratic neoliberalism at its worst (Cardullo and Kitchin, 2019). The gathering and analysis of data from the smartphones of urban dwellers has been heavily criticized as a form of "data colonialism" (Thatcher et al., 2016). The management of cities and people using the algorithmic analysis of big data is seen as replicating and deepening urban inequalities of many kinds.

It is without doubt the case that many of the phenomena described in this chapter are deeply bound into various forms of capitalist exploitation. Many of those scholars who have paid attention to the importance of urban aesthetics in recent years have felt compelled to use a rather different vocabulary to understand its power dynamics, however. An early essay by Allen (2006), for example, on a shopping mall in Berlin's Potsdamer Platz described its form of power as a kind of "seduction" rather than a deception forced on urban dwellers, against their best interests, by unscrupulous developers. Describing this as a form of "ambient power," Allen shows how the spatial design of the place and the sensory experiences of the mall appeal to the individual through an illusion of inclusion on an intimate and corporeal scale that works through the sensations that place evokes (see also Klingmann, 2007). In the case of Potsdamer Platz, this creates an environment where "power is exercised through a seductive spatial arrangement, where the experience of being in the space is itself the expression of power. Choices are restricted, options are curtailed and possibilities are closed down by degree through the forum's ambient qualities" (Allen, 2006: 454). Here power is corporeal, sensory, intimate—working through the experiences of bodies interacting with environments.

While there is a vigorous critique of the political economy of platform urbanism, and of the neoliberalization of urban governance, much less has been said about these sorts of forms of power, enacted in the sensorial corporeal grafts of urban digital embodiment. The work of Jacques Rancière is particularly appropriate to understand how power works through the new urban aesthetics. For him, power operates precisely through the everyday perception, conceptions, and lived sensory experiences. For Rancière, modern social life is organized through sensations. Rancière's (2006) own use of the term "aesthetics" often focuses on the visual arts and literature. Nonetheless, his body of work provides a useful framework for thinking about the power relations of the new urban aesthetic (see also Lindner and Sandoval, 2021). A key term in his work is "the distribution of the sensible":

> I call the distribution of the sensible the system of self-evident facts of sense perception that simultaneously discloses the existence of something in common and the delimitations that define the respective parts and positions within it [. . .] This apportionment of parts and positions is based on a distribution of spaces, times and forms of activity that determines the very manner in which something in common lends itself to participation and what ways various individuals have a part in this distribution. (Rancière, 2006: 7)

For Rancière, the "self-evident facts of sense perception" are not self-evident at all: they are precisely the working of power. What and who can be sensed—what can be seen, heard, smelled, tasted, and so on—is neither natural nor is it uniformly distributed among bodies. Some senses are represented as more highly valued than others in conceived urban space. For example, certain color palettes are preferred for building design; the smell of urine is something to be eradicated from a redeveloped neighborhood. Particular technologies are deployed to achieve this, for example the software and hardware that produces CGIs and the billboards that display them. Various "forms of perception, forms of interpretation" are allowed to some and not to others (Rancière, 2006: 80). In other words, the ways in which we perceive the world is based on a regime of socially regulated sensory frameworks. This extends an analysis of how power works in digitally mediated cities beyond the extractive processes of platform capitalism and neoliberal redevelopment, and into the realm of the sensory. In other words, it addressed the power relations inherent in the new urban aesthetic.

An example of the powerful impact of the distribution of the sensible on urban environments is Ghertner's (2015) analysis of the spatial redevelopment of millennial Delhi, where he identifies a new mode of spatial governing that has become entrenched: a "rule by aesthetics" in which modes of appearance rule which are based on sensory paradigms. Here, those in charge of the redevelopment of the city have moved away from calculative instruments based on maps, statistics, health indicators, and so on to "an aspirational vision of what the world-class urban future should look like" (Ghertner, 2015: 184) fostered by expectations set by "carefully crafted [digital] image work by real estate firms, the media, and international consultants" (Ghertner, 2015: 184). Drawing on Rancière, Ghertner discusses how this anticipated "world-class aesthetic" future is not an outcome of a discrete set of actors, but operates as a form of sensory power/knowledge which "solicits its membership in that public—a 'community of sense'—by asking potential members to partake in its ways of seeing and saying" (Ghertner, 2015: 185).

Rancière links this distribution of the sensible to the organization of space and time:

> It is a delimitation of spaces and times, of the visible and the invisible, of speech and noise, that simultaneously determines the place and the stakes of politics as a form of experience. Politics revolves around what is seen and what can be said about it, around who has the ability to see and the talent to speak, around the properties of spaces and the possibilities of time. (Rancière, 2004: 13)

The delimitation of spaces and times is clearly evident in planning, architecture, urban policies and regulations, and so on, in a number of ways. Land use zoning, the management of traffic flows, regulations about public assembly, the visualization of new public spaces, and so on all define and differentiate specific urban spaces and times and also distribute particular kinds of bodies across those spaces, for example. Moreover, who is constituted as able to contribute to the making of such distributions is also unevenly distributed: to experts, or citizens, or lawmakers, for example. These figures are audible and visible, while the voices other urban figures—the indigent, the homeless, the disabled—are heard less and acted upon much more rarely. Thus, specific practices of seeing and being seen, listening and being heard in the city are also spatially and temporally distributed in specific ways. Rancière sums up this doubled aspect of the distribution of the sensible by saying that it concerns the definition both of particular time spaces and of the bodies that can occupy them.

Many scholars suggest that the digital mediation of urban spaces is shifting distribution of the sensible. There is no doubt that contemporary usages of urban digital technologies are "inserting new scales and speeds into the warp and weft of everyday life" (McQuire, 2016: 20; and see Jansson, 2013; Fast et al., 2018): new infrastructures of the sensible are emerging, as new ways of doing things with digital devices emerge. These framings are temporal and spatial, but to address the spatial first: McQuire (2016) identifies several specific aspects of digital technologies that are part of a reorientation of urban space. For example, he discusses the constant attentiveness to content via mobile and networked digital devices, and suggests that this constant availability of many kinds of data diminishes the significance of the material built urban environment and enhances the resonance of what arrives via digital media. Social networks are one example: face-to-face interactions now happen alongside frequent online interactions with people in many different places (De Waal, 2013). And if digital media make distant people proximate, so too they bring other places into the city. Casetti (2015), in his discussion of how films are watched attentively at any time, in any place, on multiple and often mobile devices, describes one consequence as a "hypertopia," with multiple locations visible in any one place. This mix of technologies and practices thus make urban spaces multiple and "extensible" (Kitchin and Dodge, 2011: 78).

Just as digital technologies being used in ways that add particular spatialities to cities, so too they add different temporalities.

> The temporal is not a general sense of time particular to an epoch of history but a specific experience of time that is structured in specific political and economic contexts. The temporal operates as a form of social power and a type of social difference. (Sharma, 2014: 9; see also Adam, 1990)

The recent radical transformation of urban form and economy has encouraged attention to the temporal, mostly from an urban planning perspective (Crang, 2001; Degen and Lewis, 2020; Harvey, 1990; Raco et al., 2008; Soja, 1996), or looking at the economic dimensions of the 24-hour city or night-time economy (Adams et al., 2007; Roberts and Eldridge, 2012), or in relation to branding that references historical heritage (Hakala et al., 2011; Hetherington, 2013). While these studies provide important insights into how temporal power relations operate to shape the making of new urban spaces, attention also needs to be given to the temporal relations imbricated in the perception, conception, and lived aspects of the new urban aesthetic.

One example of how specific organizations of space and time allow different kinds of embodiments is Datta's (2020) examination of a smart city app to address violence against women in Delhi. While the design of the smart safety app is focused on the "instantaneous time of violence by seeking to reduce download speeds, eliminate crashes and bugs, and increase the speed of intervention" (2020: 1331), for women at the margins and living in the urban peripheries "technologies of time [are] related to use and access to urban infrastructures such as public toilets, transport, and public spaces seek to reduce travel times, increase physical and digital connectivity, and make work life more efficient" (Datta, 2020: 1331). Thus, there is a disjuncture and discrepancy between the everyday spatiotemporal needs, uses, and experiences of women within Delhi's urban infrastructure and the uses, accessibility, and working of the app. This illustrates "how a logic of speed vested in the increasing 'techno-digitization' of cities (Herzog, 1995) imagines violence against women as a set of asynchronous urban events, rather than cyclical, periodic, and socially significant time spaces of gender power inequalities" (Datta, 2020: 1320).

In the chapters that follow, we pay attention to the temporal and spatial organization of the new urban aesthetic as a way of understanding how new expressions of power are articulated through it. Each chapter explores a different version of the new urban aesthetic, and each examines how its distribution of the sensible consists of specific configurations of urban spatiality and temporality. We examine how this configuration only makes certain sensations visible, or only responds to certain kinds of bodily sensibility, or only intensifies certain bodily dispositions. Drawing on Rancière's notion of the distribution of the sensible allows the elaboration of a specific approach to aesthetic power: how aesthetics are organized by particular formations of time and space which constitute different embodied experiences. Chapter 6 will develop a critical vocabulary to describe these different distributions, and will explore in more detail how power relations in this new urban aesthetic are multiple, diverse, and contradictory (and therefore full of potential for being done differently).

6 Conclusion

Ngai (2010: 948) remarks that "our aesthetic experience is always mediated by a finite if constantly rotating repertoire of categories," and notes that these categories are both sensuous and conceptual. This chapter has addressed the aesthetic in both ways. It has described the importance of sensory experience to urban redevelopment projects and urban experience more generally; and it has offered a set of conceptual categories to address that experience. Like Ngai, the argument presented here proposes that particular aesthetic categories align with the aestheticization of commodities and cities in the current era of capitalism. We also argue that new forms of sensory urban experiencing are emerging through digital mediations that are changing how cities are felt, represented, and lived. Novel entanglements between bodies, cities, and technologies are redefining sensory urban experiences and reconfiguring spatiotemporal power relations, calling for new conceptual frameworks and analytical responses. Let us briefly recap our argument.

We started by contextualizing the emergence of the new urban aesthetic. First, we pointed to the increasing importance of branding to cities: both to entire cities and also, more relevant to our case studies in this book, to urban redevelopment projects. We suggested that much of this branding entailed efforts to choreograph and intensify bodily sensations in the city. Second, we noted how digital technologies play a vital role in both producing new urban spatialities and in enacting new forms of experience. We pointed to a number of mediations that seem particularly important to the new urban aesthetic and its sensations. For example, computer-generated images are created to both design new buildings and places and to conceive, sell, and promote the future feel of those places; images of urban neighborhoods circulate on social media platforms like Instagram; and the use of smartphone apps more generally mediate many urban experiences.

We then moved to define our preferred understanding of *aesthetics* in this book. We understand the aesthetic as the sensory relations between particular bodies and distinct material environments. These relations create the unique feel or texture of specific urban places. Drawing on the work of Böhme and others, we emphasized the staged nature of contemporary atmospheres in present-day capitalism. In the current era of aesthetic capitalism, digital technologies mediate urban aesthetics.

We also argued that it was necessary to think about how the new urban aesthetic was differentiated. We did this in four stages. The first was to emphasize that the new urban aesthetic varies between cities. The second was to emphasize different forms of aesthetic labor that experience an aesthetic. The third was to argue that urban aesthetics can be thought of in the same way as Lefebvre understood urban space more generally: as triadic. An urban aesthetic is a particular textured mix of different conceived, perceived, and lived components. Fourthly, we turned to the uneven and complex power relations that are being reconfigured as part of the new urban aesthetic. Drawing on Rancière's notion of "the distribution of the sensible" we suggested that his work provides a way of thinking about the power relations that are enacted as new urban aesthetics are experienced. It allows us to focus on the sensory; but it also encourages attention to the differential distributions of different kinds of sensibilities. How specific distributions of (particular kinds of) visuality and audibility enact power is complex, as Browne's (2015) discussion of the *un*-visibility of Black bodies exemplifies. Browne (2015) uses the term "*un*-visibility" to point to both the frequent erasure of many Black bodies from the field of visibility (not seen as [fully] human, Browne elaborates) and also to the hypervisibility of specific Black bodies (particularly when subject to the surveillance of a white gaze). This ambivalent visibility is sensory. It is also organized through particular spatialities and temporalities, as Brown's work also suggests: she is particularly attentive to the ways the surveillance of Black bodies entails the production of specific "borders, boundaries and bodies" (Browne, 2015: 16). Thus any distribution of the sensible renders some bodies more visible, more audible, more powerful in urban spaces than others, and these distributions are policed by institutions, conventions, and practices. The new urban aesthetic entrains different corporeal sensibilities differently. And given this complexity, contradiction, friction, protest, and resistance always emerge in modes both blatant and subtle (Rose et al., 2015).

Much of urban space is now "a *jointing* . . . of media and immediacy" (McQuire, 2016: 6), and in a similar fashion many urban bodies are now "a kind of graft, which is an unequivocal mark of connection and difference" between the fleshy body and virtual body (Munster, 2006: 23). The book proposes that, under aesthetic capitalism and specifically in the context of the imperative to brand urban places, those joints and grafts take powerful *aesthetic* forms. Each of our case studies describes a particular version of that form, differentiated by its types of aesthetic labor, its particular triadic texture, and its spatial and temporal distribution of the sensible. The book now turns to those case studies.

3

The Conceived Aesthetics of Urban Redevelopment

The Case of Msheireb Downtown, Qatar

Coauthored with Clare Melhuish

1 Introduction

The previous two chapters have conceptualized the new urban aesthetic as entangled with both digital technologies and urban redevelopment. This chapter begins our exploration of different examples of the new urban aesthetic with a case study of an urban redevelopment scheme which made extensive use of highly aestheticized computer-generated images (CGIs). As the previous chapter noted, CGIs are now a pervasive part of the marketing of new urban developments, large and small, in very many cities across the world. Competition for investment via branding images on a global stage has led to the trend whereby architectural images of proposed spectacular buildings or developments have become key advertising strategies for cities as "architecture shifts towards the logic of branding in its production of immaterial values of economic significance" (Arantes, 2019: 17). Architecture has become an important part of urban branding strategies in the creation of a never-ending cycle of seductive images of future buildings which circulate across the globe in magazines, exhibitions, and award ceremonies (McNeill, 2008; Ren, 2011). These "logotectures" are "now assessed according to their visual impact which reinforces the importance of the appearance of skins and surfaces . . . in a new photogenic superficiality" (Arantes, 2019: 7). Cities brand themselves with these images in order to attract attention and speculative investment, even though the projects pictured are not always commissioned.

> It is an architecture that circulates as image, and is therefore born as a figuration of itself, in a tautological circle that reduces the architectural experience to pure visuality, which is a result of a ceaseless quest for uniqueness and the "rent of form." (Arantes, 2019: 3)

This chapter explores in some detail the use of a set of forty-two digital images in the early design and branding of a large urban redevelopment project in Doha, Qatar, which had commissioned its master plan and its buildings from several world-leading architectural practices.

This set of digital images in some ways looked very similar to very many other images of such large projects—light, clean, alluring, with relaxed people doing relaxing things—even though the project developer intended them to convey a particular sense of place (see Figure 3.1). The chapter explores the sensory qualities of this aesthetic and describes it as glamorous. The chapter thus focuses directly on the role of CGIs in the aestheticization of urban change, and on a particular, glamorous instantiation of the new urban aesthetic.

So, some aspects of what we describe will be familiar to anyone with any experience of urban branding anywhere in the world. The CGIs created

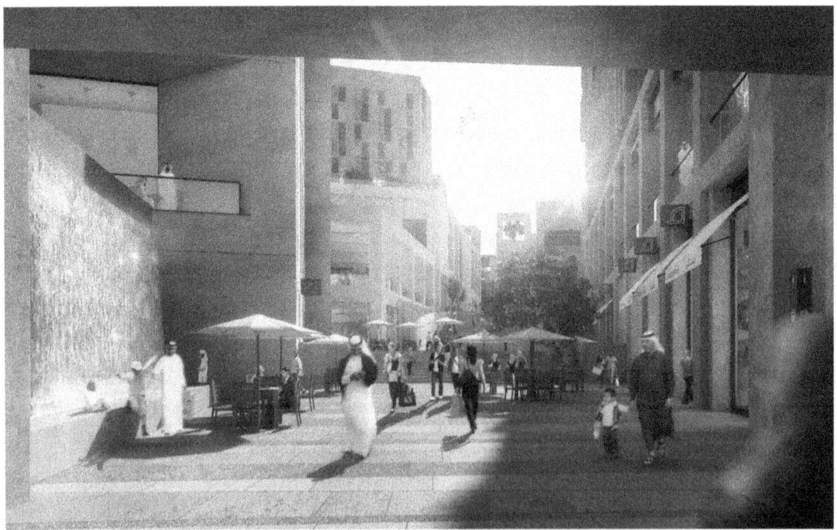

FIGURE 3.1 *A computer-generated image of a Msheireb Downtown street scene. Image copyright Gensler.*

for this project were lush, lovely images, designed initially to persuade the developer to invest in the project and then investors and families to buy its houses and apartments and rent its shops and hotels. They stage beautiful buildings, plazas, and malls inhabited by happy people doing fun, leisurely things. They were designed in the offices of well-known architecture or architectural visualization companies at considerable expense, and much of this professional aesthetic labor was expended in efforts to visualize the right kind of atmosphere. Indeed, the developer wanted images (and buildings) that were somewhat different from generic "logotecture." According to its website, the mandate of the project was based on "the principles of progress and tradition, freedom and responsibility, new and old cultures, and innovation and stability," which in design terms meant a combination of traditional Qatari architectural forms and modern sustainable buildings to induce "the rebirth of a neighbourhood and the reawakening of an architectural heritage."[1] As the chapter will demonstrate, visualizing that rebirth and reawakening was not straightforward. The glamorous urban aesthetic is "an outcome of performative acts of manipulation that aim to captivate and engage audiences in sensory ways" (Huopalainen, 2019: 333). It entailed extensive working and reworking of image after image before the developer and architects were satisfied. Visualizing a new place and its values takes a lot of aesthetic labor.

In the studios and meeting rooms of architects, visualizers, the project's development management team and their client, the images were discussed and altered again and again to ensure that they pictured certain things in particular ways. The previous chapter noted how Böhme (2017) describes this as the staging of urban life: the various elements of these images were combined, modified, and recombined until they felt appropriately atmospheric. The digitality of the images meant they could be revised repeatedly (Rose, 2016), until they made appropriate reference to various aspects of the project which were highly valued by the architects and developer. The chapter thus acknowledges the perceived aspects of these CGIs but emphasizes their conceived quality. As the previous chapter noted, conceived space is constituted when "knowledge of [urban] material reality is comprehended essentially through thought, as *res cogito*, literally 'thought things'" (Soja, 1996: 79). The CGIs discussed in this chapter are one example of such "thought things." While such conceived things are often suggested to be rational abstractions, in the current moment, these highly crafted CGIs represent Msheireb Downtown's qualities in very sensuous ways. In the new urban aesthetic, even conceived spaces can be sensory.

---

[1] www.msheireb.com

The chapter begins by introducing its case study in more detail. It then explores the sensations of the Msheireb Downtown CGIs. Specifically, the chapter describes these CGIs as *glamorous*. Glamour is a particular technology of visual allure (Thrift, 2008); it is one way in which urban atmospheres can be seductive (Allen, 2006). The chapter identifies the specific sensorial techniques of glamour that pull the viewer into the CGIs' atmospheres. The sensations evoked by the spaces visualized in these images have a particular spatial pull, and their implied temporalities also entice the viewer in. The final section considers some of the abstractions that this glamorous conceived urban space represents. The previous chapter discussed our various strategies for embedding differentiation into our account of the new urban aesthetic, including working with specific examples of that aesthetic and understanding it as triadic. The final section touches briefly on the latter point, but pays more extended attention to an aspect of these conceived images: the way that the abstractions to which they refer are multiple and sometimes contradictory. Conceived space is itself textured. These CGIs are thus an excellent case study for examples of the differentiations which always disrupt any new urban aesthetic.

2 The Case Study: Introducing Msheireb Downtown

Our case study in this chapter is a $5.5 billion project on a 31-hectare site in the historic center of Doha, Qatar, currently called Msheireb Downtown (Msheireb means "a place to drink water" in Arabic). It is a complex urban transformation project which has been driven by the rhetoric of urban placemaking and where a variety of digital visualizations have been central in visioning, producing, and communicating the project. The neighborhood to be redeveloped was a run-down area of low-rise homes and businesses occupied since the 1970s by mainly migrant workers from Kerala in South India, which has been transformed into a mixed-use zone of public spaces, residential, hotel, and office accommodation as well as a museum area and shopping center.

The design of Msheireb Downtown began in 2008 with an international competition to select a group of architects "that could interpret in modern terms, a vernacular that could bridge the gap between the Doha of the past and the Doha of tomorrow".[2] Construction started in 2010 and was scheduled to finish in 2018, although at the time of writing parts of the site are still under construction. The concept master plan was produced by the US-based

[2] www.msheireb.com

company AECOM, and AECOM drew up the detailed master plan with global consultants Arup and London architects Allies and Morrison. There are a further range of sitewide and executive consultants with specific responsibilities, coordinated by the Master Development Consultants (Msheireb Downtown) team, and including executive consultants, executive architects, and landscape architects, among others. Via the competition, nine design architects were chosen to design the hundred or so buildings on the site.

The developer—the architects' client—is Msheireb Properties, a company owned by the Qatar Foundation. The Qatar Foundation is a public-private organization set up by the Emir of Qatar and Sheikha Moza bint Nasser. It consists of around fifty organizations in education, research, and community development and gives this redevelopment project some specific values and agendas. In the context of this project, the Qatar Foundation was concerned to develop a distinctively Qatari neighborhood that would "foster the reawakening of an architectural heritage in danger of extinction" (Msheireb Downtown Doha—Mandate 2020). As the project website further notes, "three years of intensive research was invested in ensuring that the blueprint of Msheireb Downtown Doha integrated the true spirit and aesthetics of Qatari architecture with modern, highly functional and sustainable development." All the architects were asked to work to a specified design code of Qatari architectural design elements while also designing a model of urban living that currently does not exist in Doha in order "to re-energise the core of the city and create a hub of activity, where people return once again to live, work, shop and spend time with family and friends."[3] The project thus aspired to distance itself from a "starchitect," Western model of development (Melhuish et al., 2016). It also intended to initiate a new form of urban living:

> Although the Qatar Foundation undoubtedly sees the project as an act of cultural philanthropy, it is nonetheless asking to make a profit. [. . .] it wants to build a model that other developers will seek to emulate. [. . .] to create a market for a form of urbanity for which there is currently no demand [in the Gulf]. (Woodman, 2012)

As we will see, the Qatar Foundation's desire for a distinctively Qatari urbanism promoting a "clear cultural identity expressed through planning and architecture" (Law and Underwood, 2012: 131) was one of the key values that drove this redevelopment.

In other ways though, notwithstanding what one architect called the "Qatari character mission," the project exemplifies "the new downtown [as]

---

[3] www.msheireb.com

a redesign of the urban centre" (Rotenberg, 2012: 30) materialized through a "transnational architectural production" process (Ren, 2011: 5): all but one of the eventual architects were based in London, the other in Barcelona. At the heart of the development, typical of Western models of urban development, is a large public square, addressed by a Cultural Forum at one end, and a luxury hotel at the other (see Figure 3.2). The square is lined by shops, restaurants, and outdoor seating along each side, and has a sophisticated climate control system (including overhead canopy, spilled air, and cooled water) designed to reduce outdoor summer temperatures by 10 degrees and ensure this new public space is also usable for as much of the year as possible. The development also features a series of grand new archive buildings facing the government palace to the north; on the other side is a large open prayer ground, and a cultural quarter of reconstructed "heritage houses" converted into museums. An extensive indoor shopping galleria, luxury shopping quarter, hotels, and offices comprise a substantial part of the development, along with apartments and townhouses for segregated Qatari and foreign residents to rent on the north-west section of the site.

One of the earliest tasks of the Msheireb Downtown master planners was to design a suite of CGIs to sell their vision of the development to Msheireb Properties and the Qatar Foundation, to persuade them to invest in the redevelopment project. As one of the architects in the Msheireb project reflected, "you can draw elevations and plans and sections and CAD drawings and they have to be right, but at the end of the day I suppose what convinces the client is the sales images." Those CGIs were then used in a whole range

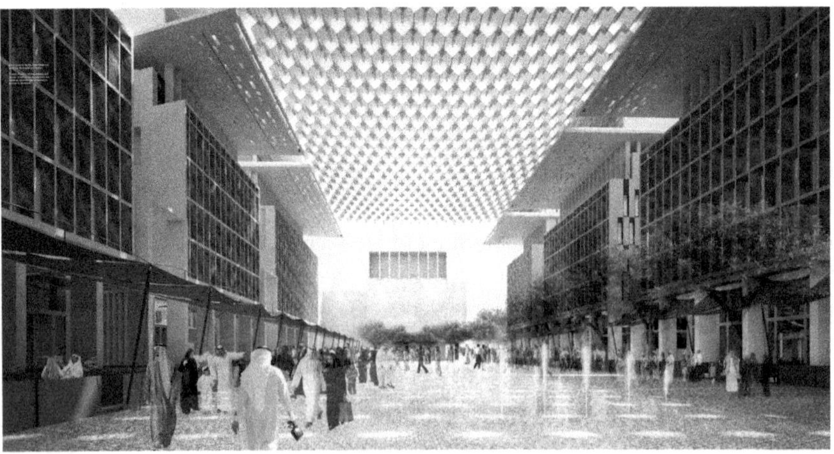

FIGURE 3.2 *A computer-generated image of the public square in Msheireb Downtown. Image copyright Mossessian Architecture.*

of other marketing activities that were all part of the branding the Msheireb Downtown development as an attractive place to invest, play, shop, and live.

However, the use of CGIs in the Msheireb Downtown project was somewhat unique in that its design as well as its branding relied heavily on CGIs of what it would look and feel like when it was complete. CGIs are created for many redevelopment projects in order to convince a whole range of urban actors that a proposed development project is desirable: developers at meetings where visions for new projects are pitched; policy makers and community representatives in planning meetings; the residents, workers, and tourists driving or strolling past the building site; other developers at real estate fairs; investors, or private clients who might buy a flat or a hotel building or rent a retail space. In the case of Msheireb Downtown, CGIs were also part of the design process. Digital visualizing softwares of various kinds have become a routine part of architectural practice, partly as design tools and partly because they allow architects' clients to visualize the physical design of buildings and public spaces and to suggest changes and reach an agreed design. However, those tend to be two somewhat distinct uses of digital visualizing technologies, and indeed often use different softwares. In this case, as we will see, the design and the branding process were very closely aligned.

The Msheireb Downtown manager recalled that initially there were "around twenty images . . . more describing the mood of the whole development." At the early stages of the development, their role was to picture the kind of urban life the development would host. An architect described those first CGIs like this:

> those were the ones that were not necessarily showing all the architecture, describing the architecture. They were more describing the mood of the whole development. So, it was not about "that's this building, that's that building." They were showing more that's the life of the street, that's what you feel by taking the journey through the development.

In describing the "mood" and the "feeling" of the development, these CGIs are clearly part of the move described in the previous chapter toward branding places through aesthetic sensations and sensibilities. As another architect said, at this stage images are suggestive: "some of it is very evocative, but it has to be at that stage. I think it goes back to the notion of who is receiving the imagery and how they understand it. Those images are not meant to be sturdy imagery, they are evocative and they emphasize things." A Special Design Review meeting in Doha in 2012 finalized forty-two such images to visualize the masterplan. Since then those images have been circulated in a variety of formats and among different audiences both within and beyond Qatar. These included staff and visitors to the client's and architects' offices, visitors to the permanent exhibition at the Msheireb Enrichment Centre on a barge moored off West Bay, international property fairs such as MIPIM in Cannes, and social media sites including the client's own, international newspapers and magazines, as well as passers-by in cars and on foot along the perimeter of the site in Doha, where the images were mounted on billboards, and contractors and construction workers within the site who used the images as reference points. They were also used as design tools (see Figure 3.3). The architects found that CGIs allowed the developer to understand the broader urban development better as "the visuals were the means to communicate to someone who is not an architect." They were used in design review meetings with the developer to visualize the design of individual buildings and their compliance with the project's design code, as well as how they would look alongside each other. This chapter focuses on those forty-two CGIs chosen in 2012 to show what the development would look like when it was finished.

The production of the forty-two Msheireb Downtown CGIs followed a fairly consistent process (in Chapter 6 we elaborate in more detail on this aesthetic labor). Two digital visualization studios were employed to do this work, one in London and the other in Liverpool. A CGI started life as a wireframe model

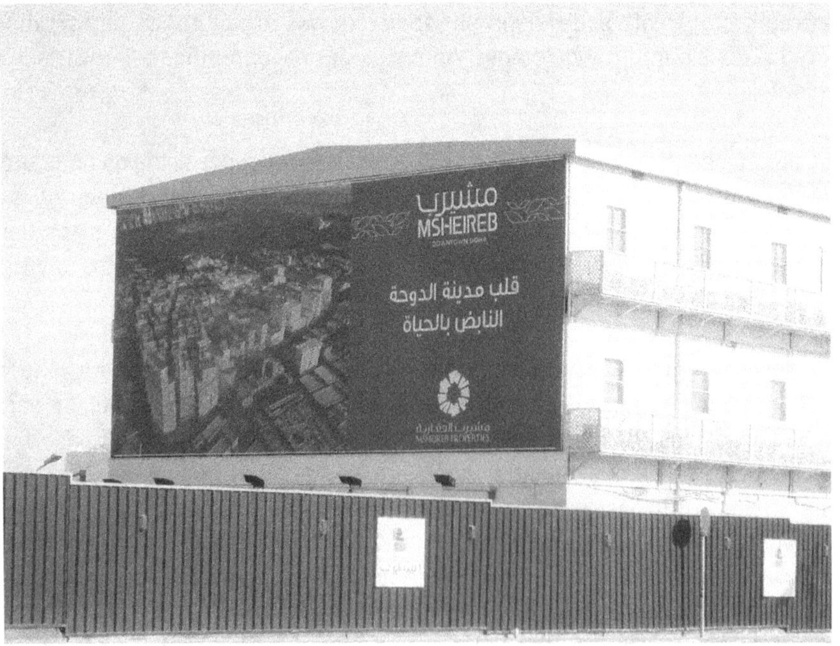

FIGURE 3.3 *A hoarding advertising Msheireb Downtown on the site of the development, photographed in May 2012. Image copyright Claire Melhuish.*

generated by the Computer Assisted Design (CAD) software used by a team of architects to design a building. That file was then imported into a visualization software package, usually 3DS Max. Visualizers then situate the model in the 3DS Max's embedded Global Positioning System: it is vital that the visualizations are realistic in terms of the sun's angle, so the model is sited using the GPS, the visualizer specifies time of year and day and the software generates the correct angle of sunlight that will hit the building in that place at that time. In CGIs with several buildings, several models are imported. The visualizer then strips out a lot of the architects' design details, after which layers of colors, materials, and textures are added to the model's surfaces, which can still be manipulated in three dimensions on the visualizer's screen. At certain points, the file is sent away to be "rendered": that is, to be converted from a working image to something that looks more like a photograph. The file then returns to the visualizer for more work. Once the building itself is looking good in 3DS Max, the file is imported into another software package, usually Photoshop, to be worked up still further as a two-dimensional image. It is at this point that elements of the staged scene that are difficult to produce convincingly using only digital production tools are pasted in: trees and, as we will discuss in Section 4 of this chapter, people (see Figure 3.4).

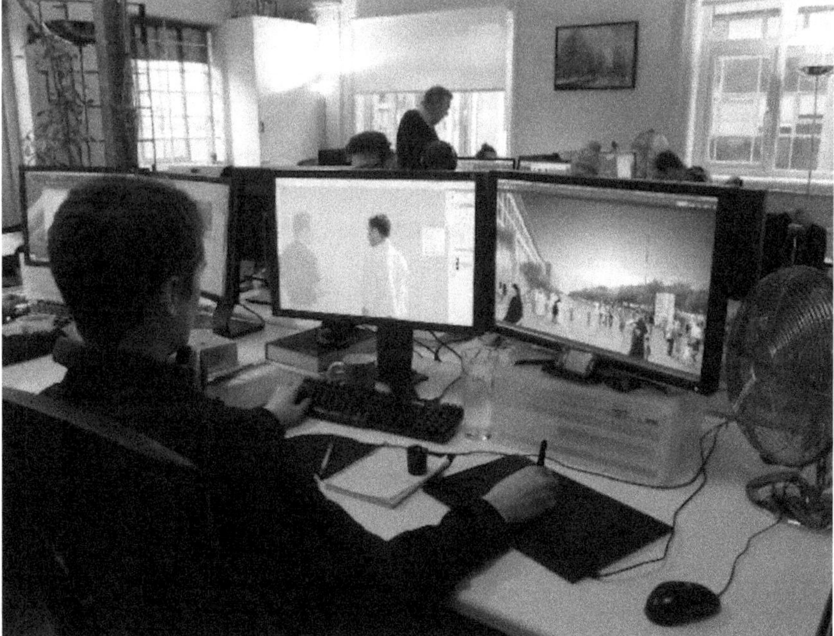

FIGURE 3.4 *A visualizer pasting photographs of figures into a computer-generated Msheireb Downtown street. Image copyright Claire Melhuish.*

At various points in this process, the visualizer will have sent the file back to the architect for viewing and feedback on their work. In our case study, there was also the project's Architectural Language Advisor (ALA). The ALA was a partner at the architecture firm which developed the project's master plan. His role was to oversee the visualization process. He reviewed all the CGIs and offered guidance and concrete advice on both the accuracy of the images and the mood they evoked. Indeed, the ALA wrote a guide for all the project visualizers specifying ten ways to generate such feelings, which he called "magic moments." The ALA also signed off the CGIs before they were shown to the client. This was often by writing comments on printed versions of the draft CGIs, scanning the printouts and returning them to the visualizer. Hence, every CGI has very many iterations, as it gets modified by different people in the production process. The ALA was a particularly enthusiastic advocate of the project's CGIs as a means of cohering the project's different components into, as he said, "something one wants on the cover of a magazine." Here, then, we have a developer of a large-scale urban redevelopment project investing significant time and money in creating a suite of CGIs that stage—to use Böhme's (2017) term—"how that place will look and feel" when it is finished, to quote a Msheireb Downtown manager.

The CGIs of Msheireb Downtown were continually required to evoke a distinctive and powerful aesthetic. Indeed, they are strikingly atmospheric. They picture the Msheireb Downtown development either in daylight in a slightly washed-out palette of beige, blue, and green, or at night when the pale buildings gleam warmly and many kinds of lights glow into the ink-dark sky. There are other qualities to the images too. They show a clean and orderly environment of regular, rectilinear, almost classical stone-clad buildings, on a grid-based plan of open boulevards and vistas, interlaced with intimate *siqats* or alleys like those in the fast-disappearing old city of Doha. The buildings have traditional Qatari details, such as *badgir* (wind chimneys) or *liwan* (covered terrace or open-fronted room). In some images, overhangs and colonnades create integrated, shaded outdoor spaces, filled by informal groups of Qataris and expatriates enjoying themselves in dappled light; others picture the dazzling play of lights at night (Figure 3.5); others show streets and malls full of luxury brands and five star hotels.

As the Msheireb Downtown redevelopment project continued through more phases of more detailed design and then construction, this set of CGIs continued to play a number of roles in the development process, often being modified as they did so: as advertisements for the development but also for the architects and visualizers; as part of discussions about Msheireb Downtown's design; they were even used as evidence by the architects that they had completed design tasks so that their fees could be paid. This then

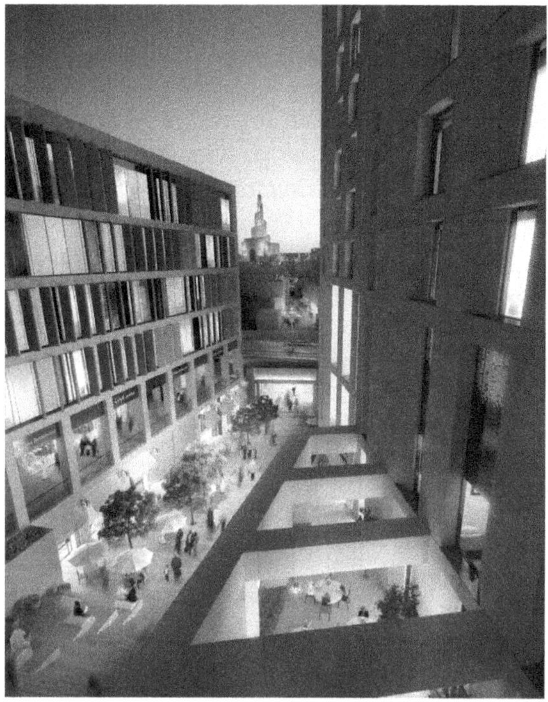

FIGURE 3.5 *A computer-generated image of the Msheireb Downtown project at dusk. Image copyright Gensler.*

was "an architecture that circulates as image" (Arantes, 2019: 3), but the images' role in the design process meant that as images, they were the focus of many discussions about what the project was trying to achieve. Hence this chapter approaches the CGIs as constituting *conceived* urban spaces. Among other things, they represent particular design features and how they would appear when they were built; they refer to a particular place; and they reference the developer's particular vision of Qatariness. They staged, costumed, and intensified what life would be like in Msheireb Downtown when it was complete (Böhme, 2003). And they did so with a version of the new urban aesthetic that we will describe as glamorous.

3 Spatializing Urban Glamour: "Something One Wants on the Cover of a Magazine"

The previous chapter remarked that for several commentators on urban atmospheres, atmospheres associated with urban redevelopment projects are seductive (or they try to be, at least, a point we will return to in Section

5 of this chapter). Glamour too has been understood as a particular kind of sensory atmosphere which extends from things and captivates in particular ways (Brown, 2009). Glamour shows in luminous appearances that offer immersion without friction or effort in a world without troubles. Glamour is visual: color, light, and texture all contribute (Brown, 2009; Hearn and Banet-Weiser, 2020; Thrift, 2008). Glamour stages something desirable and apparently real, such that the viewer is caught up with the image and what it offers. "By binding image and desire, [glamour] gives us pleasure. . . . It leads us to feel that the life we dream of exists, and to desire it even more" (Postrel, 2013: 6). As one visualizer put it, rather prosaically, "I suppose that's why visualisers put people on benches drinking coffee, happy scenes"; he catches the sense that glamour is not just a picture of glamorous things but also a feeling. For Postrel, picturing glamour must generate "enough familiarity to engage the imagination, allowing scope for the viewer's own fantasies" (Postrel, 2013: 20), and must remain a little vague since a "glamorous image appeals to our desires without becoming explicit, lest too much information breaks the spell" (Postrel, 2013: 20). Defined in this way, glamour saturates all the Msheireb Downtown CGIs, so we describe their particular configuration of the new urban aesthetic as glamorous. In this section we explore its specific allure and how it is articulated through the sensory organization of the CGIs' spaces.

The forty-two Msheireb Downtown CGIs visualize not individual buildings but a range of street scenes. In contrast to the distancing "hero shots" of flagship architecture which are common in many advertisements, the Msheireb project visualizations show clusters of buildings surrounding public spaces. Many elements of these scenes are glamorous in the sense that they show beautiful buildings; happy, stylish, and relaxed people; clear skies; children playing; high-end shops and hotels; green trees; almost no traffic; and many other elements of what is conventionally understood as a luxurious and stylish urban experience. This is definitely a world without troubles, full of ALA's "magic moments."

The CGIs allow their viewer to imagine themselves present in that world by offering a certain kind of spatial seduction. Most of the Msheireb CGI's have been produced to create close views carefully composed to suggest an embodied experience of being immersed in the space—walking or cycling perhaps—not gazing at it from a distance as a spectator or consumer. They have been carefully directed at different stages by visualizers and the project's ALA to achieve a sensuous and atmospheric quality that will draw the viewer in by appealing to their sensations. A glowing dusk sky, a sea breeze or a fountain splashing are meticulously visualized to evoke memorable feelings. It was not unusual to have five or more versions per viewpoint drafted. Eye-level, as opposed to bird's eye, camera angles are chosen and the picture is then developed to give a particular feel of the area as experienced by a particular viewer. Indeed, one of the ALA's favorite images is taken from a child's perspective:

> [. . .] this is a nice view, and this view is from child's eye height, and if you think of being a photographer, you often do this, everyone does that [crouching down], you get a completely different perspective, it's remarkably different.

Adopting this sense of first-person perspectives is an important way in which the viewer is drawn into the CGI, helping him or her to imagine sensorially what it would feel like being there in the present moment (see Figure 3.6). Böhme (2013) argues that the changing perspectives that are invoked through the visual or physical movement or journey are what produces an overall atmospheric bodily sensation of a place, an illusion of being a participant through being able to experience the space. Analyzing the effect visualizations have on their own ongoing bodily dispositions and orientations. Bissell and Fuller (2017: 2487) also argue that a viewer's immersion into an image "at the level of direct sense" is accentuated by a depiction of speed and movement that generates "sensory pleasures" (see Figure 3.6).

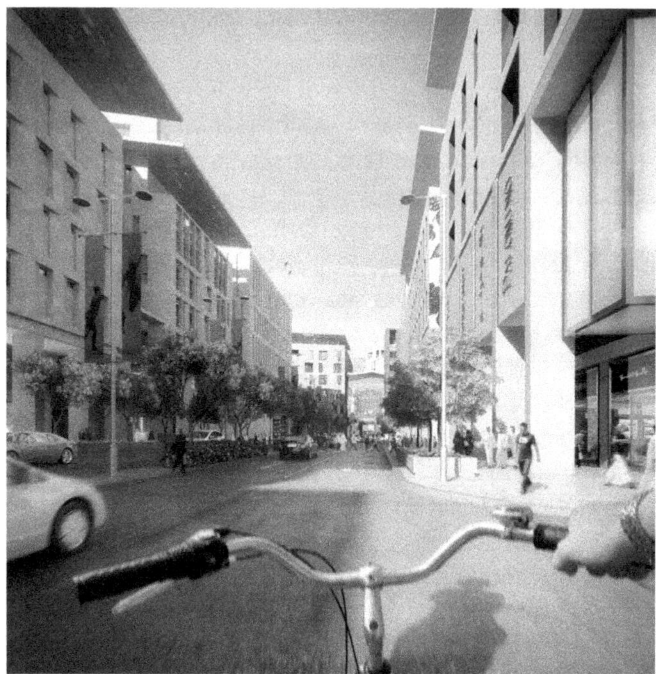

FIGURE 3.6 *Cycling through Msheireb Downtown. Image copyright Gensler.*

Chapter 2 discussed how in digitally mediated cities, bodies were grafted between their corporeality and their digital mediation. This chapter has suggested that these CGIs enact that grafting between CGI and viewer as a relation of glamour, and thus far has emphasized one of its conceived aspects: the representations of leisured urban life. But that urban experience is clearly also pictured through perceived space. Perceived urban aesthetics refers to the more "material, sensuous dimensions of the media; the very stuff in terms of tools and infrastructures for mediation that make up our everyday environments" (Jansson, 2013: 282). The glow of light that radiates across these urban scenes and turns them into envelopes of glamour also extends out to their viewers, and this is part of the expressive power of brands in aesthetic capitalism which we discussed in Chapter 2. The viewing of images is always a coproduction between the viewer's own experiences and the reading of an image, hence "in viewing the image we can draw upon our embodied experience to feel [what is depicted] proprioceptively even though we are not feeling an actual [object]" (Ash, 2009, p. 13 quoted in Kinsley, 2010: 2783). Hence, to enroll the viewer's imagination and anticipation of the new environments, the ALA was also keen to produce images that would invoke the whole embodied experience of the new development, such as the sounds of the new Msheireb Downtown. He wanted rustling trees, the play of water, and the bell of a tram to be audible, somehow, from the visualizations. These were part of his typology of "magic moments": visual elements which he insisted gave life to these staged images. A magic moment was "a slice of life which carries a story and also a resonance" which would add integrity to the picture such as "a bike, blurred, in motion," or "a child, running or smiling, towards the camera." As the ALA further explained, "you don't always want a balloon and an ice-ream and a smile, but they're great, smiles are good!."

The viewer is also lured into these CGIs not only one by one, but also because they offer the "journey through the site" that one architect noted. CGI by CGI, the viewer is taken from view to view of public and semi-public outdoor spaces between buildings, enlivened by different types of everyday activity. Most people or groups of people are depicted in motion, often walking, some blurring through their movement across the picture. Similar to how Ash describes videogames, we can see that those making CGIs "actively manipulate spatiotemporal aspects of the [CGI] to produce positively affective encounters for users (by which I mean encounters which increase the body's capacity to act and produce associated positive senses of intensity)" (Ash, 2010: 654). There is a feeling of constant activity, of movement, so that the images never feel static, fixed, immobile, or frozen. They are—to quote Thrift (2008) on colorful glamour—dynamic and alive.

The CGIs are glamorous, then, in that their spatial organization invites the viewer to immerse themselves into their sensory world. One of the most striking visual aspects of the glamorous spaces of the Doha visualizations is the quality of the light that they picture, and this also helps to allure and immerse. The light congeals between the things in the images: it is what radiates to form the atmosphere between them (Böhme, 2013, 2017: 61). There was a lot of concern that the lighting of the buildings, streets, and squares of Msheireb Downtown should be accurate. Indeed, as the previous section noted, one of the first things that a visualizer does after importing a CAD model into their visualizing software is to locate and orient the building accurately using the software's GPS in order for the software to calculate the angle and position of the sunlight that would be falling on that building at that moment, in that location. Doha light emanates from featureless skies and produces high contrast in the visual field. As a result, an architect told us, "the play of shadows in the Middle East is very important." However, visualizers were concerned with a lot more than being accurate. The point of the animated sunlight tracking studies created by one visualizer, for example, was to ensure that shadows in the CGIs would be "poetic and dappled" because, according to the ALA's guide for visualizers, "shadow can be emotive." Trees can "shimmer and glisten," there should be "grazing sunlight on facades," he added. In several interior daytime images, sunbeams are deliberately emphasized (known as "god rays") and give the picture and the individuals depicted a glowing atmosphere. This is what one visualizer described as "the dreamy feel" that visualizers aimed for in these CGIs; they were both accurate but with that slight vagueness that Postrel (2013) identifies as crucial to glamour.

But that work to create light effects was also related to the Qatar Foundation's mission to picture a new form of distinctively Qatari urbanism. The light had to look Qatari. Not only did the direction of light have to be accurate but so too did its location. This was described in a number of ways. Doha light is bright white with a slight haze: "almost humid," said the ALA. Even the more dramatic images—in which shafts of light plunge into darker enclosed spaces, for example—were somehow visually muted by a slight bloom and dust, and a limited color range of sandy beige buildings, washed-out blue skies, and pale green trees. These various efforts to make the light effects in the CGIs "accurate" indicate the ways in which these images' sensoriality are part of the conceived aspects of this urban redevelopment project. The notion of a light that is distinctively Qatari makes images of the light a kind of abstraction. It gives a particular interpretation to a quality of light: it is hazy because it is Qatari. The sensory qualities of the CGIs are assumed to refer to other things: to symbolize or represent them. Indeed, in her discussion of glamour, Brown (2009) argues that it too is a kind of abstraction. It is a thing translated into an idea. Glamour is "a form of non-verbal rhetoric, which moves and persuades not through words but through images, concepts, and totems" (Hearn and Banet-Weiser, 2020; Postrel, 2013: 6), and thus aligns with the conceived qualities of this new urban aesthetic.

The wide and subtle range of light effects in these images contributes in very large part to their glamorous atmosphere, then. Light glows and beams, flows and scintillates, glances and shadows, bursts and dims; "[w]hat has been produced is what might be called a spirited sense of colour, which goes beyond colour as such in that it incorporates tactile movement—wheelings and pivotings and splicings—into its effect" (Thrift, 2012: 154). It is also an effect best seen on screens. Many of these images were printed, to be displayed on hoardings, for example, or in promotional literature. But their atmosphere was crafted onscreen and is best seen on screen, and the screen itself adds its own allure. As Dorrian (2008) explains:

> [The sun] does not really cast light onto virtual objects but rather is used to calibrate the way in which the emission of light directly to the eye is modulated across the screen. Light is not cast upon, but emerges out of the virtual object, which it at the same time constitutes. Under these conditions, even when rendered with degrees of transparency, the virtual object seems taut, more than present. (Dorrian, 2008: 47)

Night shots tend to play with artificial lighting and depict mostly lit-up streets, shops but also fireworks, evoking celebratory moments. The outcome is a "splintered, refracted city compromised of a constantly shifting pattern of ambient textures, ephemeral shimmering shapes, glittering colours, and incandescent lights" (Edensor, 2012: 1107). Similar to Edensor's description of the Blackpool illuminations, we witness a luminescent staging of space, where the images try to draw the viewer in through staging a sparkling, "jewel-like" intensity, where "crucial to the hyperdefinition of the object is the scintillation and emissivity of the image on the computer screen" (Dorrian, 2008: 47) (Figure 3.7).

These forty-two CGIs thus have a distinctive digital aesthetic which we call glamorous: it is desirable and alluring. It is both perceived and conceived. Life in Msheireb Downtown looks gorgeous: Qatari, leisured, and beautiful. Some of this is visual; there is that distinctive restrained color palette, for example, and a particular luminosity. But much is spatial. The extensive range of light effects radiate between and among and from the different surface in and of the digital image. Viewers are positioned bodily in intimate relation to these spaces, often, with an unusual point of view, and are invited to move from one place in Msheireb Downtown to another as if on a journey. It was this glamorous feeling that the ALA worked to achieve, even though this desired quality of an image was difficult to prescribe verbally. For example,

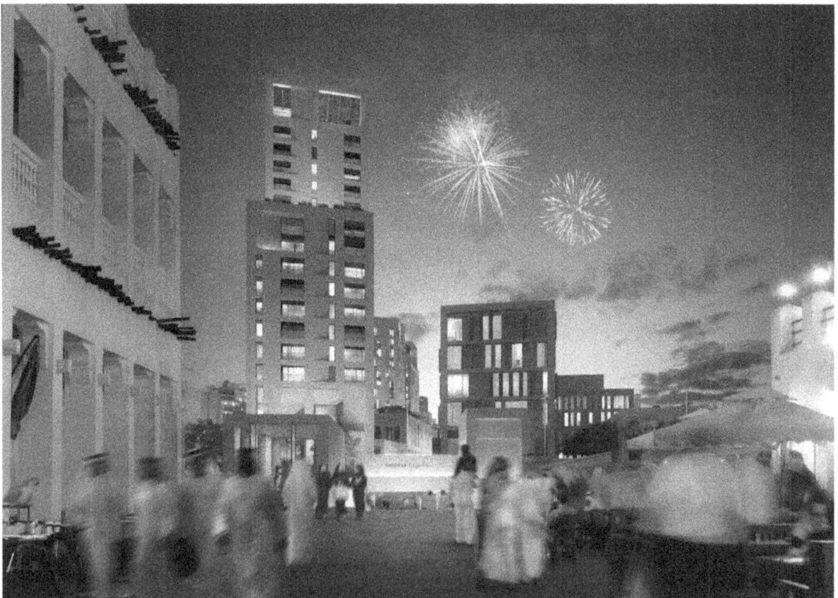

FIGURE 3.7 *A computer-generated image of the Msheireb Downtown project at night. Image copyright Gensler.*

he instructed visualizers to use the softwares' focus, mist, and blur tools but to avoid a "cosy glow." The goal was for "all of these things comes together":

> Just like the architectural photographer waits for that moment when someone walks past the camera, and the image becomes a front cover image—because it's blurred, so whether it's movement blur or out of focus or graininess or memorable moment or foreground, or camera setup, and I'm obsessed with camera setup, all of these things come together to make something one wants on the cover of a magazine.

For a CGI to make it onto the front cover of a magazine sums up the themes of this section. Not only would it have to be visually striking, but it also has to be attractive enough to make someone pick up the magazine (and buy it) for pleasure. It has to draw a viewer in. Allure, pleasure—and profit.

4 The Temporalities of Urban Glamour: "The Key to the Image Is That You Are Telling a Story"

The glamorous allure of the Msheireb Downtown CGIs also weaves its spell on the viewer through particular forms of temporal organization. Picturing an unbuilt neighborhood like Msheireb Downtown is about anticipating a specific kind of urban place, and the previous section focused on the visual and spatial qualities of the glamorous CGIs created as part of the Msheireb Downtown project. But, picturing an unbuilt neighborhood like Msheireb Downtown is also about anticipating a specific kind of urban future through the power of evocating sensations. In this glamorous version of the new urban aesthetic, that future must be enticing and desirable: it must effortlessly allure its viewers into a better, more pleasurable world. This section shows that constructing two kinds of temporal experiences in relation to that future is as important as the spaces of its luscious, glowing glamour.

The first temporality is one of transformation. CGIs are visualizations of urban environments not yet built, of potential urban futures. Planning is a deeply temporal activity as it is "the transition over time from current states to desired ones" (Abram and Wezkalnys, 2011: 3–4). Much planning practice consists of preparing for future activities by trying to organize, predict, and manage future spatial forms. As urban space is highly unpredictable and messy such strategies, plans, or future projections tend to "tame urban complexity" (Hoch, 2009). Implicit in this future scenario building are the promises for a better material order, implying that the already present needs improvement: "plans can be constructed to avoid undesirable futures, to make desired forecasts come true, or to create new, more desirable futures" (Myers and Kitsuse, 2000: 223). This was certainly the case for the Msheireb Downtown development. The new scheme would replace an area of low-rise buildings, subdivided, and occupied by migrant workers. What was regarded by the developer as an ethnicized landscape of disrepair and poor living standards would be replaced by a new contemporary Qatari landscape that would foster a leisured and wealthy lifestyle appropriate for rich and well-traveled Qataris. It is very evident here that the CGIs, as conceived representations of this area of Doha, were valuing certain visions of that area over others.

Representations of urban futures should be understood as performative interventions into practices of urban change. As they are made, so certain kinds of futures are imagined and others are rendered impossible. This is done through two main features in urban development (cf Anderson, 2010): first, calculated forecasts and projections of, for example, future economic growth, environmental sustainability, demographic change, and estimated future visitor numbers; and second, through the construction of scenarios, visioning, and backcasting which aims to provide assurances to investors and engage the general public on an affective level with what the future will feel and look like. Thus "images are a key technique through which specific futures are enacted in the present" (Bissell and Fuller, 2017: 2478). Indeed, particular future visions always provide rationales for action in the present (Anderson, 2010). Future landscapes need not only be communicated effectively but, more importantly, visualized convincingly. Because the future is the "realm of the 'not yet,'" and thus "not accessible to the senses" (Adam, 2008: 6), it needs to be evoked "as if" it was real (Anderson, 2010: 785). This does not mean that futures are visualized completely accurately. Indeed, for these CGIs to be convincingly glamorous, it is important that they are not too realistic, as the previous section noted. This perhaps accounts for the oddly clean quality of the Msheireb Downtown's CGI scenes: buildings are new and spotless, despite the hazy dust of Doha.

The spotless and unaged buildings of Msheireb Downtown are of the future, then, but they are also oddly out of time. The CGIs do not reference the transformation of this part of Doha as a long-term process, and the Msheireb Downtown landscape does not seem to age. Everything in Msheireb Downtown seems set in the ever present and entirely new: clean, undamaged, perfect. It is as if these glamorous spaces appear ready-formed. The CGIs appear to picture a conceived space which shows that "the represented urban spatial organisation can be built or evaluated as a physical situation after a given visual representation" (Paklone, 2011: 155). The CGIs are simply abstractions of a future urban reality. The transformed urban space will follow on from its image. In his discussion of futurities, Anderson suggests that statements about the future like these CGIs "function through a circularity, in that statements disclose a set of relations between past, present and future *and* self-authenticate those relations" (2010: 778–9). Etymologically, glamour is related to the notion of a magical spell. It seems the spell the CGIs are casting is an unavoidable and instantaneous jump forward into a future timeless moment: they visualize an effortless transformation, and effortlessness is a key part of glamour (Thrift, 2008).

The second temporality that structures these CGIs is the feeling of living in these spaces now. To achieve this sense of in-the-moment inhabitation, visualizers and architects repeatedly described the glamorous aesthetic as "the story," or, as the ALA told us, "storyboards are very important." This aesthetic story consists of focusing on everyday practices, almost banal moments of the urban present, yet evoking them as smooth occurrences with no conflict or sensory experiences that could disrupt the glamour of the place. Hence, we do not see rubbish or rubbish bins that could evoke smells, no dogs or pigeons that could foul on the clean surfaces, or even no dirt or aging textures that could suggest a challenge to the hygienic, clean version of the future.

As the architectural critic Till explains, of all modes of communication used in architecture, storytelling is the most productive, as "stories collapse the barriers between expert architect and non-expert client and user" (2009: 114). As he further elaborates, because stories are founded in everyday experiences—of both the client and architect—the envisaged future environments cannot be impossibly idealistic or come across as too glamourized, but knowledge and imagination are shared and externalized. Thus, the need of the aesthetic story to focus on the banal everyday and to be conveyed through the image is crucial to provide the link between the spatial and temporal gap, and between the future conceived in the CGI and the moment the client views the image. In the case of the Msheireb Downtown project, it was challenging because very few of the architects and visualizers involved ever visited Doha or met their client. As this visualizer explains:

> It is quite a big problem that you are not able to be in the room and to talk through what you are exploring with the client. Because it is about narrative and telling a story—that's what all these images are about, nothing else really. The key to the image is that you are telling a story, whether the story is: "This is what this building will look like on this particular day with this particular lighting condition and that's how the interior behaves and how the light will fall." Or whether your story is "This is the garden area that Qatari locals will go with their families." And you are more interested in the occupation of the space rather than the materiality. So, it involves an explanation that goes with the image as well.

This is a temporal version of the spatial immersion in the glamorous enveloping offered by the CGIs that the previous section described: "the role of the architect becomes to understand and draw out the spatial implications of urban storytelling" (Till, 2009: 114).

The CGIs therefore have to demonstrate the feel of future places and buildings and to suggest prospective embodied interactions by anticipating the sensations these places will generate. The introduction to this chapter quoted one architect describing the CGIs as "not about 'that's this building, that's that building'. They were showing more that's the life of the street, that's what you feel by taking the journey through the development." Kinsley (2010) describes the elicitation of such sensations as "future-oriented embodied attunements," and it is another example of the interplay between the conceived and the perceived in these CGIs. He discussed this in relation to the production of embodied anticipation in video representations of technological futures, and shows how technological futures are mediated through a bodily comprehension of the viewer, familiarity with similar technologies and a bodily and sensory attunement to potential technological experiences: "it is through these somatic markers brought about in the performance of viewing, between the representational construct of the image and the viewer's body, that we are materially enrolled into anticipation" (Kinsley, 2010: 2784). To put it simply, the experience of viewing depictions of technological or urban futures mediates the viewer's embodied experience in the present, their past memories and future hopes onto the anticipated feel of the future development. Each image entrains the present body of the viewer in an unfolding story of a particular urban future.

This section has suggested that the glamorous conceived urban aesthetic is underpinned by two temporal dimensions. First, a temporality of future urban transformation which is evocative rather than realistic, so that it remains desirable, and almost atemporal in that it is always already there, new, never aging or progressing, or developing through time. Second, a temporality that evokes the everyday feel of encountering these places without any interruption or disturbance through a smooth unfolding story. Both temporalities seduce the viewer into Msheireb Downtown: they are glamorous techniques of allure.

5 The Conceived New Urban Aesthetic: When Glamour Goes Wrong

As several commentators have remarked, glamour is a tricky sensation to stage (Brown, 2009; Hearn and Banet-Weiser, 2020). Paradoxically, while glamour exudes effortlessness, its creation demands "meticulous selection and control" (Thrift, 2008: 15), exemplified in the extensive aesthetic labor expended on the CGIs discussed here. This section explores some specific examples of that labor to demonstrate how the glamour of these CGIs was not always achieved. It begins by pointing briefly to the triadic quality of their aesthetics and in particular to different perceived relations to the CGIs. It then explores how the abstractions of their conceived spaces did not always align.

So far this chapter has focused on the viewer's sensory engagement with the CGIs. But the visualizers' embodied labor also engaged with their glamorous urban aesthetic via the interfaces of their computer screen and mouse. As Chapter 6 will describe, visualizers worked long hours sitting in offices staring at screens, working with 3DS Max to create the required images. That did not always situate those visualizers as seduced and allured by these images. On the contrary, their embodied engagement with stuff and tools was often routinized and boring and, as deadlines loomed and schedules became frantic, frustrated, and exhausted.

The conceived aspects of these images were also multiple. The dominant value represented in the CGIs was a seductive future of intimate urban spaces in a distinctively Qatari location. However, there were also other kinds of abstractions present in the CGIs. There were references to other organizations of space, for example. As we saw in Section 2 of this chapter, the CGIs start out as a version of CAD models which plan how the building will eventually be built. A simplified wireframe is generated for importing into the visualization software, direct from the same model that will produce the actual building.

We also noted how the buildings were aligned using GPS coordinates. This was crucial since Msheireb Downtown was about combining lots of different buildings into a new urban neighborhood. Early in the design process, the EC produced a document of guidelines that all the 3DS Max models created as part of the design process were obliged to follow. Its aim was "to deliver consistent electronic deliverables." The document did a number of things, including instructing visualizers to use the coordinate template provided by the EC; specifying acceptable file formats; describing the required conventions for naming each file, each material, and each layer within the 3DS Max model; noting the geometry to be used (for example, "all objects will have wielded [*sic*] vertices," "no double geometry is allowed"); and insisting on the inclusion of "all native package files" in the deliverable. In this way, the CGIs also reference a set of rather different spatial coordinates than the ones discussed so far. As well as an alluring representation of future urban life, they were also meant to be geometrically accurate.

This was the same coordinate system that the visualizers used to create the correct position and angle of the sun. This concern for the accuracy of the CGIs' light was utilized in the building site as images of what the eventual buildings would "really" look like. A number were printed out life-size and mounted on a series of hoardings on the building site, where they were used as background to evaluate different landscaping materials (see Figure 3.8). Here, the CGIs were substituting—apparently usefully—the actual eventual buildings. However, the accuracy of this sunlight could also generate discussion among the various architects. As one visualizer explained to us when discussing how to negotiate architects' expectations of what to depict:

> You then say to the architect: "This will be a Doha sun. This date at this time." But then the architect will still go: "Whoa, the stone is not shiny enough." If we turn round and go: "Well, the stone wouldn't be shiny at that time of the day," they'll still stamp their feet and say they want the stone to be shiny. So, you can't win to be honest. There's certain things which you can bend the rules on, but changing the sun probably isn't one of them.

Here the visualizer suggests that the accuracy of the sun's angle and strength generated by the software is challenged by the architect's understanding of what a stone should look like in the CGI. Quite what representations constituted accurate time and space in the CGIs was not always settled.

Another aspect of the CGIs conceived abstractions which generated some friction was the developer's desire that the CGIs had to appear Qatari, which the chapter has already noted in the context of the project's architectural design language and the aesthetics of the CGI's light. Also, the developer

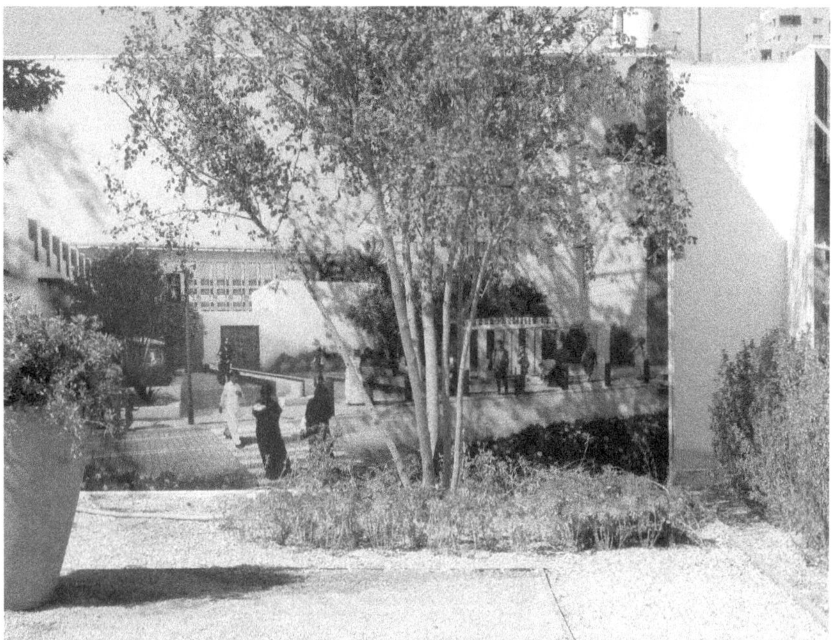

FIGURE 3.8 *A computer-generated image being used to assess hard landscaping and planting on the Msheireb Downtown site. Image copyright Claire Melhuish.*

wanted the people in the CGIs to look Qatari. Usually, digital visualizers working on architectural computer-generated images will add human figures to their visualization at a very late stage in the development of the image, often as the "final polish" is added in Photoshop, to quote a visualizer (Houdart, 2008). Photographs of people, and often of other very complex items like trees, are sourced either from the visualizer's personal archive or from online image banks, and are pasted into the image. As the chapter has already noted, the redevelopment project was backed by the Qatar Foundation as part of a cultural project contributing toward a modern and forward-looking Qatari identity; another major motivation for the redevelopment of the Msheireb site was to bring Qatari families back into the city center. Therefore, the Qatar Foundation insisted that the images had to show recognizably Qatari people, and especially Qatari families. Pictured bodies were therefore crucial to the success of these images, in the effort to sell the project to the client by showing an appropriate balance between global cachet and local distinctiveness.

This was not clear to the visualizers early on in the project, and an early animation created for Msheireb Properties "took a hammering" (Msheireb Downtown manager) from the client because it did not look sufficiently Qatari. Subsequently, much more effort went into producing appropriate images inhabited by the right bodies. Once it became clear that the client had specific expectations about wanting Qatari-looking people in the CGIs, visualizers were forced to look beyond their usual image banks to find figures to insert into the Msheireb Downtown CGIs, since those sources held very few, if any, properly Qatari-looking figures. One tactic was to search the online photo-sharing website Flickr for holiday snaps taken in Qatar that they could cannibalize. Another was to create their own. Thus, in a warehouse in London on a cold November morning in 2012, in a huge room with white walls and floor, a photoshoot was organized by the studio who had been tasked to rework some of the forty-two CGIs. Three photographers were taking shots of individuals who had been recruited through a film extras agency in London: with printouts of the CGIs at hand, a Msheireb Downtown visualizer was there too, asking the extras to act as if they were strolling and chatting in the public spaces the CGIs were picturing. These photographs of people were then masked and pasted into the revised CGIs. The extras were costumed in what the visualizers thought were Qatari-looking clothes (actually bought from Asian stores in Southall, a London suburb). The representation of human figures in the CGIs turned out to be no less problematic than the abstraction of time and space.

A number of points can be made from this episode. For one, it underlines our analysis of the Msheireb Downtown CGIs as conceived spaces, in that the bodies in these images have to persuasively refer to the notion of Qatari identity. There had to be families and there had to be women in hijab but also men and women in Western dress, as well as men in Qatari dress. As long as they looked right, appearing to refer to Qatari practices, their actual relation to Qatar was unimportant. Another point is that the CGIs make no reference to bodies other than these wealthy Qatari citizens: for example, the workers who until their eviction in 2009 had lived in the neighborhood, many of whom were working to build Msheireb Downtown (an erasure also generated by the transformational temporal jump into the future of the finished development pictured by the images). These CGIs make only certain kinds of bodies visible, rendered in the glamorous atmosphere of this timeless urban space, and we will discuss this in more detail in Chapter 6. A further key point is that this is one example of many, where different actors wanted the images to refer appropriately to different things.

Indeed there were many other examples of discussion, debate, and disagreement about what the CGIs should look like on the basis of what the images represented. This could be about representing the buildings, or Doha, as we have seen. It could also be about different kinds of professional expertise. Throughout the Downtown Doha project, there were many tensions between the client, architects, master planners, the ALA, and visualizers, who all wanted different kinds of images which would do different things by looking different. Neither the architects nor the ALA liked the marketing "hero shot"; the architects tended to be much more fascinated by "the backstreet world." Indeed, the architects and the master planners were often uncomfortable with the sorts of CGIs made for the client, both as part of the pitch and as part of the design process. They are "one sided," according to a master planner: good at "engaging people" but not so useful for design work because they often show "things that lead to inappropriate discussions." Another architect explained, "people tend not to look at the architecture but at the image and make decisions based on liking or not liking the image." This was a problem for the architects and master planners, partly because they understood their professional practice as designing three-dimensional buildings, not two-dimensional surfaces. Visualizers on the other hand precisely emphasized their communication skills. One pointed to the tension between his desire to produce a beautiful, well-composed image to enable the client's engagement with the design, and the architects' desire to show the details of their buildings, which led to his views on how the image should look being "slightly hijacked." The expectations of architects in using CGI's as sales pitch, the visualizers role in creating an accurate image and the clients' needs for an attractive and comprehensible picture. A visualizer commented on the work that then happened to reconcile these different requirements: "the visualiser needs to understand exactly what the designer is trying to achieve with that image . . . you need to give some context and that takes a lot of time. It's quite a slow process and it's not going to happen on its own." As a consequence, the images are labored over again and again.

The conceived aspects of the CGIs of the Msheireb Downtown development represent many things; they are abstractions of temporal and spatial coordinate systems, sunlight, a place, and various kinds of professional expertise, several of which may contradict each other. What this section has emphasized is that, as conceived abstractions, there is always the potential for representations to multiply or fail.

6 Conclusion: Texturizing Glamour

This chapter has focused on just one example of one of the many intersections between urban branding and digital technologies: the use of CGIs to design and to sell urban redevelopment projects. We explored how glamour can be regarded as a key feature of Msheireb Downtown's new urban aesthetic. A glamorous new urban aesthetic is an aesthetic of luminous surfaces picturing a desirable, trouble-free future urban life. We argued that this sensation of glamour is evoked through the specific spatial and temporal organizations of the urban development depicted in the CGIs. Urban glamour is achieved spatially by immersing the viewer into the depicted urban spaces through pedestrian-level viewpoints enveloped in lovely light, and by creating a constant sense of movement through these spaces. Temporally, Msheireb Downtown's CGIs offer a smooth and seductive urban future, ageless and decay-free, into which the viewer is sensorially enticed through an unfolding story.

All of this is achieved through computer-generated images. In their discussion of digital images, Hoelzl and Marie describe what they see as their distinguishing features:

> As a program, the image, while still appearing as a geometrical projection on our screens, is inextricably mixed up with the data (physical and digital) and the continuous processing of data. What was supposed to be a solid representation of a solid world based on the sound principle of geometric projection (our operational mode for centuries), a *hard* image as it were, is revealed to be something totally different, ubiquitous, infinitely adaptable and adaptive and intrinsically merged with software: a *softimage*. (Hoelzl and Marie, 2015: 7)

There are moments when this adaptability directly mediated the process of designing Msheireb Downtown. For example, relatively basic, pre-Photoshop CGIs can be quickly changed, which allowed architects to experiment with different volumes and materials. As one architect said, CGIs (made in Sketchup in this case) are a "quicker means of showing the massing, changing massing. We used to have a top floor overhang; when we removed it, what did it look like?" These basic CGIs were also shown to the client to see if a new idea was worth pursuing. In these instances, a building could be entirely reconfigured in moments.

More generally though, this chapter could be framed as an exploration of how a softimage gets hardened up, as it were: how its infinite adaptability is stabilized. Particular versions of space and time are especially important to this process. Some of these versions of time and space were more or less uncontroversial among the actors in our case study: the use of a consistent GPS coordinate system, for example, to allow the integration of separate buildings into street scenes, or of certain CGIs that are seen as visually accurate enough to allow decisions about what landscaping materials to use with them. Representations of conceived spaces the CGIs were solidified, if you like, by being calibrated via systems of spatial coordinates and geometries that have long been used as part of architectural and planning processes. Digital visualization tools were used to mimic "solid representations of a solid world based on the sound principle of geometric projection."

Indeed, many of the conceived elements of the CGIs could be understood as efforts to stabilize the image by making it refer to other things in "accurate" or "realistic" kinds of ways, and some of these could be a little more contentious: how sunlight might change colors, for example. Others were much more unstable. As we have shown, the CGIs seen to represent Qatari identity were particularly problematic. Design features, for example, could be modified but showing Doha's dusty light and its future inhabitants as Qatari was more challenging for the visualizers and they needed considerable work.

The conceived aspects of these images can then be understood as multiple and sometimes contradictory. And as textured images, there was something more directly sensory which the images needed to generate: a sensation of glamour that was nonrepresentational. This is the new urban aesthetic as triadic, as perceived spaces intertwine with conceived spaces. Perceived elements entail a more directly sensory engagement with the images, which is hard to describe but critical for glamour to be staged effectively. "I don't know how you describe it but you can see it," said one architect, while another said that their visualizer was so good because of "a sensitivity" that allowed them to invoke a particular expressive feel through the use of colors. As Thrift notes, in an experience economy, "[t]he practices of worlding demand the use of a much greater sensory palette in order to produce ambience as well as message [. . .] The intention is to produce atmospheres—tropic or frozen, cramped or spacious, busy or still but atmospheres all the same" (Thrift, 2012: 153).

The new urban aesthetic of glamour is perceived as well as conceived. The Msheireb Downtown CGIs picture an untroubled world, and their glamorous spatiality and timelessness renders them particularly alluring, immersing the viewer in their hazy light and future moment, the atmospheric light binding the pictured bodies and the viewing bodies to the buildings-to-come. But as we have demonstrated in this chapter, not only is a new urban aesthetic textured, as conceived and perceived elements play with and against each other, but there are differences embedded in each of those qualities too, as conceived representations are challenged and perceived relations grip or not. The new urban aesthetic is differentiated all the way down.

4

The Perceived Aesthetics of Digital Urbanism

Feeling Digital Embodiment in Smart Milton Keynes

1 Introduction

This chapter explores another manifestation of the new urban aesthetic. It works with the case of the so-called smart city. The phrase "smart city" emerged in the mid-1990s and was consolidated in 2011 when IBM trademarked the phrase "Smarter Cities" as a way to sell its data management software to cities. A smart city integrates different kinds of digital data about the city, from different sources and in real time, in order to improve city functioning (for overviews, see Aurigi and Willis, 2020; Hollands, 2008; Luque-Ayala and Marvin, 2020; Willis and Aurigi, 2018). City authorities and commercial providers suggest that smart policies and technologies can enhance the environmental sustainability of cities, for example, by enabling the more efficient use of resources, especially energy and water; or that urban economic growth can be increased by innovating new products and markets based on digital data; or that populations can be surveilled more extensively; or that "smart citizens" can participate more effectively in urban governance and social or entrepreneurial innovation. Smart has also become part of the place branding discussed in Chapter 2. To be seen as a smart city is to be seen as modern, clean, green, and forward-thinking. For city leaders in Barcelona, Manchester, Amsterdam, Singapore, and many more—including our last case study Doha which was awarded the "Smart City" award in 2018—their ranking in league tables of smart cities is part of their reputational brand and a vital part of their efforts to compete as a global city.

So the chapter begins by looking at how cities are staged as smart. The previous chapter argued that staging and storytelling was an important way in which specific kinds of urban futures were visualized by developers and architects. Storytelling is no less central to the proponents of smart cities, who also see smart as the future direction of city development. Many accounts of smart cities have explored the extensive work that is done to make the smart city credible to city councils and residents (Datta, 2018; Halegoua, 2019; Kitchin, Claudio, et al., 2017; Luque-Ayala and Marvin, 2020; McNeill, 2015; Sadowski and Bendor, 2018; Söderström et al., 2014; Vanolo, 2014). Corporations, startups, thought leaders, computer scientists, policy makers, elected representatives, campaigners: all these actors tell stories about smart cities. This chapter examines the stories staged in visualizations generated by companies and cities promoting smart tech and smart cities. As with any kind of urban branding now, digital images are central to much of this storytelling, and the chapter begins by exploring a range of examples, including from the city that is this chapter's case study, Milton Keynes in the UK.

These images of smart urban life may be digital, but they are not glamorous. The chapter progresses our discussion of the new urban aesthetic by arguing that these stories and images manifest a different version of the new urban aesthetic. They do not picture urban life in the same way that the CGIs of Msheireb Downtown did. Instead of an aesthetic of glamour, this chapter identifies a version of the new urban aesthetic that is all about flow: the flow of digital data through a city. As Luque-Ayala and Marvin (2020: 7) remark, in smart cities, "constant information flow becomes the new nature of the city, the milieu that needs to be created," and this is the milieu visualized in the promotional materials examined here. This version of the new urban aesthetic valorizes that flow. This chapter's description of a new urban aesthetic of flow exemplifies our argument that there are different configurations of the new urban aesthetic.

In that sense, the images of smart cities discussed here are conceived: they represent a particular vision of digitally mediated city life. The chapter's case study though suggests that the urban aesthetic of flow especially generates a particularly *perceived* urban space. Our case study city, Milton Keynes, has for some years been the site of a large number of different kinds of smart urban projects involving many different kinds of local organizations, including its city council, the local universities, large corporations, a national innovation hub, community organizations, small startups, and more. In the second part of this chapter, we take a look across this smart city scene and discuss a range of smartphone apps designed with a number of these local organizations, examining both the stories told about them and their imbrication inflows of smart urban data.

Apps installed on smartphones are an important means by which data in a smart city is generated (Dieter et al., 2019; Schwanen, 2015; Thatcher, 2017; White, 2016). Apps are also particularly interesting ways to think through the sensory bodily implications of the new urban aesthetic of flow. Installed on smartphones, accompanying people at every moment of their everyday life, their use mundane, repetitive, and habitual, apps can be regarded as "functional and sensorial prostheses" for very many bodies (Srnicek, 2014: 83). The smartphone screen becomes the interface between the corporeal body and the smart city. Apps are thus examples of the kind of digital embodiment discussed in Chapter 2. They exemplify what Munster (2006: 23) calls "a kind of graft, which is an unequivocal mark of connection and difference" between the fleshy body and virtual body. Lupton (2016, 2019) has paid sustained attention to the sensorial dimensions of interactions with data, exploring how digital devices like apps turn embodied sensory experiences into data which then mediate feeling. She discusses self-tracking apps in particular, but all smartphone apps are an example of "the more material, sensuous dimensions of the media; the very stuff in terms of tools and infrastructures for mediation that make up our everyday environments" (Jansson, 2013: 282). Apps generate important perceived elements of the flowing new urban aesthetic, and as a tool and an infrastructure of mediation embedded in the everyday, apps are part of a smart city's expressive infrastructure (Thrift, 2012).

After an introduction to smart cities in general and Milton Keynes (MK) in particular in Section 2, this chapter discusses smart city storytelling as part of urban branding, including in MK. The discussion addresses the textured quality of this new urban aesthetic. To describe a new urban aesthetic as textured is to emphasize that it has aspects that are perceived, conceived, and lived, and that these may not neatly align (Jansson, 2013). The chapter's discussion of MK's branding explores how the new urban aesthetic of smart city flow conceives both a particular and an abstract or generic city. The chapter then turns to the group of smart city apps designed as part of smart city projects in MK and explores how they generate and circulate data among their various users. The chapter pays particular attention to the embodied, perceived sensation of movement which the apps' designs were all intended to induce, and specifies the graft between the digital device, the data it generates, and the body. Finally, the chapter returns to visualizations of smart cities and suggests that they also constantly flow. Not only do they circulate widely, but they are also themselves animated, as are their viewers in consequence (and see Brighenti and Pavoni, 2020). Rather than being enveloped by a seductive glamorous glow, this smart new urban aesthetic induces a smooth mobility. The chapter concludes by exploring the relation between these conceived and perceived smart spaces.

2 The Case Study: Milton Keynes, Smart City

MK is the last of UK's postwar new towns. It has had stories told about it from its founding in 1968, starting with the efforts of its development corporation to attract businesses and residents to move to the new city in the 1970s and 1980s using billboard, newspaper, and television advertising campaigns, to its current lively presence on social media (in Chapter 5 we will say more about urban branding on social media). It has also consistently experimented with new urban technologies: for example, it was the site of the UK's first solar-powered housing in 1972. It is now in the top half-dozen of the UK's smart cities (Huawei, 2017; Smart Cities UK, 2020). MK council officers and elected representatives attend smart city national and international expos and events to showcase its ongoing smart activities and also to pitch itself as a city in which future smart experimentation is welcomed. While MK has, or had, some large-scale and long-term smart projects, many more are small and relatively short-lived. In that sense, it is similar to many—perhaps most—other smart cities (Cowley and Caprotti, 2019). As a relatively small city trying to be, and be seen as, a smart city (in fact, MK is not officially a city, it is a town), it is an interesting case study to explore what might be called "ordinary smart," rather than the efforts of much larger and better-known cities like London, Doha, Manchester, Barcelona, or Amsterdam.

So what is a "smart city"? "The dominant vision of a smart city," write Luque-Ayala and Marvin (2020: 13), "is one of a digitally enhanced urbanity that combines intelligent infrastructure, high-tech urban development, the digital economy, and electronically enabled forms of citizenship." The gathering and analysis of data to improve life in cities has a long history (Barns, 2020; Halpern, 2015; Krivý, 2016; Luque-Ayala and Marvin, 2020; Mattern, 2016). In a "smart city," the city authority (often in partnership with private companies) collects, integrates, and analyses big, geolocated, real-time digital data, in order to manage its city more efficiently and sustainably. Recent scholarship is also focusing on a much wider range of organizations which use digital devices of many kinds to harvest data from urban environments, however, analyzing how not only efficiencies but also profits are sought through the extraction, circulation, transformation, commodification, integration, storage, and reuse of data. This broader context of the "colonization of the [urban] lifeworld through the commodification and extraction of personal information as data" (Thatcher et al., 2016: 992) is part of "platform capitalism," in which financial value is extracted from the collection and analysis of data (Srnicek, 2016). Discussions of big data in smart cities have thus been joined by discussions of "platform urbanism" (see, e.g., Barns, 2020; Fields et al., 2020; Lee et al., 2020; Leszczynski, 2019; Richardson, 2020b).

Considerable attention has been given to the data infrastructure required by both smart and platform urbanism in order to transmute the city via the use of data into "a calculative and logistical enterprise" (Luque-Ayala and Marvin, 2020: 21). Luque-Ayala and Marvin (2020), for example, explore computational packages and control centers as well as the development of specific systems, platforms, and networks. Others have discussed the design and delivery of dashboards which offer online open access to big urban data (Kitchin et al., 2016; Mattern, 2015) and have unpicked how sensors and smartphones generate very particular kinds of data about a city (Gabrys et al., 2016; Kitchin, Lauriault et al., 2017; Shelton, 2017; Thatcher, 2017). Notwithstanding Kitchin's (2014: 1) warning that data-driven cities risk being "buggy, brittle and hackable," several commentators have argued that the ubiquity of digital data is creating new kinds of flowing urban space (for an early statement, see Castells, 1996; and Luque-Ayala and Marvin, 2020; McQuire, 2016).

Thus defined, the smart city has been heavily criticized from a number of angles. There is a significant literature which positions smart cities as part of a global—if differentiated—intensification of neoliberal forms of urbanism after the 2008 crash (Cardullo and Kitchin, 2019; Greenfield, 2013; Grossi and Pianezzi, 2017; Hollands, 2008; Kitchin, 2014; Pollio, 2016; Vanolo, 2014) and which focuses on the economic policies and ideologies of smart cities. It argues that they are neoliberal because economic growth is prioritized over, for example, environmental sustainability or social justice (Joss et al., 2017; Luque-Ayala and Neves Maia, 2018); or because the market is presented as the solution to urban governance such that entrepreneurialism, innovation, outsourcing, and profit are normalized; or because "the social contract is replaced by the corporate contract" (Sadowski and Pasquale, 2015). Concerns have also been raised about surveillance and privacy in smart cities (Iveson and Maalsen, 2018; McNeill, 2016; Sadowski and Pasquale, 2015), and about the relatively weak form of agency epitomized in the notion of the "smart citizen" who inhabits the smart city (Calzada, 2018; Cardullo and Kitchin, 2018; Ho, 2017; Shelton and Lodato, 2019). In Kitchin's (2014) influential statement, all these criticisms culminate in the overarching accusation that smart cities are neoliberal technocracies which aim to replace democratic debate with data-driven management (and see Luque-Ayala and Marvin, 2020).

Many of these criticisms have also been made of the related phenomenon of "platform urbanism," as Chapter 1 noted. Attention has focused on how the data collated by platforms enacts "an asymmetric power relationship in which individuals are dispossessed of the data they generate in their day-to-day lives" (Thatcher et al., 2016: 990). Thatcher et al. (2016) point not only to how profit is made from individuals' data by corporations and not individuals; they also show how such data quantifies and surveys urban life as "previously private times and places are commodified and privatized as a new terrain for capital investment and exchange" (Thatcher et al., 2016: 991; and see, e.g., Sadowski and Pasquale, 2015; Schindler and Marvin, 2018). It has been noted that platform users do not properly profit from their sharing of data (Barns, 2020; Graham, 2020; Richardson, 2020a). And significant attention has been given to the algorithmic processing of big platform data, which highlights that as algorithms are trained or learn from existing data, various kinds of social relations become embedded in their working. Noble's (2018) discussion of Google's search engine is exemplary here; and Benjamin (2019) has discussed how racism is enacted in facial recognition software (often a flagship technology in smart city efforts to increase urban security).

Many of these criticisms of smart cities and platform urbanism are justified, especially when large corporations are involved in smart city projects (see, e.g., McNeill, 2015; Wiig, 2016). However, platform urbanism and smart cities are not quite the same thing and many smart cities do not operate on a platform model (Sadowski, 2020). Rather than projects run with, or by, large software and hardware corporations, many cities are going smart in a more piecemeal and less centralized manner (Cowley and Caprotti, 2019; Taylor and While, 2017). There are now many different kinds of smart city technologies. Much smart city activity is funded by a range of corporate and public organizations in the form of short-term experiments or research projects, often awarded in competition with other cities. This is another reason why there is lot of visualizing of smart cities, not only as part of urban branding but also as a way of heightening a city's visibility and credibility in these competitions. Combinations of these technologies operate under various modes of organization, ownership, accountability, and mission. The line between "smart" and an experiment in digitally mediated urbanism is not always clear. This is what has been called the "urban laboratory" model of a smart city, and while it works with the creation, merger, and analysis of digital data, those processes are neither as large nor as integrated as platform urbanism (Karvonen and van Heur, 2014).

MK falls into this latter type of smart city. It has no control center, nor does its council have a smart city strategy group or even a chief information officer or equivalent. Instead, it has had for the past few years an enthusiastic director of strategy and council leader, who have brought a wide range of smart organizations and projects into the city primarily to support its economic growth. While the first mention of "smart" in MK appears to be in a council transport strategy review in 2007, the development of MK as a smart city began around 2012, based on the director's insight that smart technologies needed to be tested on the ground somewhere. As he says:

> Often enterprises, organisations that were innovating and developing new approaches needed to do that somewhere and that gave me a bit of insight into something that has stood us well over time, which is thinking about our place as a place which can be welcoming and supportive and attractive and enticing for those sorts of projects.

In that sense, smart in MK is "organic," to quote the director, formed of whatever projects can be persuaded to land in, or emerge from, the city (although transport has emerged as a broad priority in recent years).

The "urban lab MK" strategy has produced a number of successful bids for smart city projects. Examples over the past decade include an £8 million grant for the deployment of charging infrastructure for electric vehicles, a £13 million smart grids trial, and £150 million for the Transport Systems Catapult (TSC), which until 2019 was a national innovation hub for transport headquartered in MK. Another major smart project coordinated by the TSC involved trialing connected and autonomous vehicles (CAVs) in MK. Another project, MK:Smart, was led by computer scientists at the Open University with a number of partners between 2014 and 2017: the local council, other universities, water and electricity companies, and telecommunications providers (Valdez et al., 2018). Further significant investment arrived in 2018 when sensors were installed to monitor and manage traffic flow and car parking spaces across the city. Both the TSC and MK:Smart facilitated many smaller projects, often in partnership with other funders. As a result, MK is a smart city full of relatively short-term projects with significant amounts of public and civic as well as corporate involvement, and with a wide range of smart stakeholders including charities, voluntary sector organizations, and local campaigners as well as MK:Smart and TSC partners.

As noted earlier, most of the attention given to these and other kinds of smart city projects has focused on their data infrastructure. Our discussion focuses instead on the expressive infrastructure of smart cities: their new urban aesthetic. None of the smart city projects in MK had the resource to pay for the highly finished CGIs of the sort discussed in the previous chapter. Nonetheless, some rather less polished CGIs have been created, most projects had websites and Twitter feeds, their apps have visual interfaces of course, and some commissioned promotional videos. These were often designed by a local digital visualization company—let's call it DVC—which creates CGIs, animations, and interactive apps, its website explains. It made a number of videos with various partners in city projects located in the city, and also a recent promotional video for the council's future growth plans. The next section looks at these and other efforts at smart branding, and begins to sketch the new urban aesthetics of flow.

3 Conceiving Smart Cities: Storytelling about Smart MK

The idea of the smart city as expressed by its advocates—corporate or civic, large or small—is seductive. Smart technologies are repeatedly framed as solving urban problems—and who does not want that? However, the previous section noted the many criticisms that have been made about the particular kinds of solutions (to specific kinds of problems) proffered by smart technologies. Most critiques focus on questions of justice, democracy, accountability and rights. Contributing to this body of work, we want to examine the seductiveness of the smart city and in particular the seduction of the new urban aesthetic of flow. This section explores how it works as the conceived aspect of smart city branding. The section also suggests that, as the previous chapter noted, the conceived values embedded in that branding are multiple.

The hardware and software companies that sell smart city technologies put a lot of work into advertising their products, and visualize them on their YouTube channels, in advertisements, through their Twitter feeds and at expos. Siemens has even built an entire exhibition space in London's Docklands to promote smart sustainable urbanism. Cities also create videos and animations, policy documents, and Twitter campaigns. Smart cities are pictured in photorealistic computer-generated images and digital animations; graphics explain smart tech; photographs provide evidence of smart activity (or its failure). Such images display many kinds of content in diverse styles and genres; they are materialized in specific formats; and they are viewed in many different situations.

Some patterns are discernible however. Notably, cities are presented as places where more and more humans will live in the future, but also where there are very many complex problems to be solved (Halegoua, 2019; Rose, 2018). "Smart" is then presented as the solution to those problems, via the analysis of big data. That data is persistently visualized as flowing through the city. This is an abstraction in the sense that data flows are not visible to the human eye, so these images must imagine them; they must be symbolized, and in these images this is done most often by glowing networks of blue lines pulsing through city landscapes. Data and its circulation is repeatedly visualized as an airy light show: in a 2015 video made by Innovate UK to showcase its investment in a range of smart city projects, including MK, these networks flash up like fireworks over video footage of urban landscapes, or look like some kind of webbing that appears over city streets and as a background to photographs phone apps, laptops, and desktops (see Figure 4.1). Smart city promotional images also tend to have a distinct color range, with the blue that is used to picture digital data and digital networks very prominent,

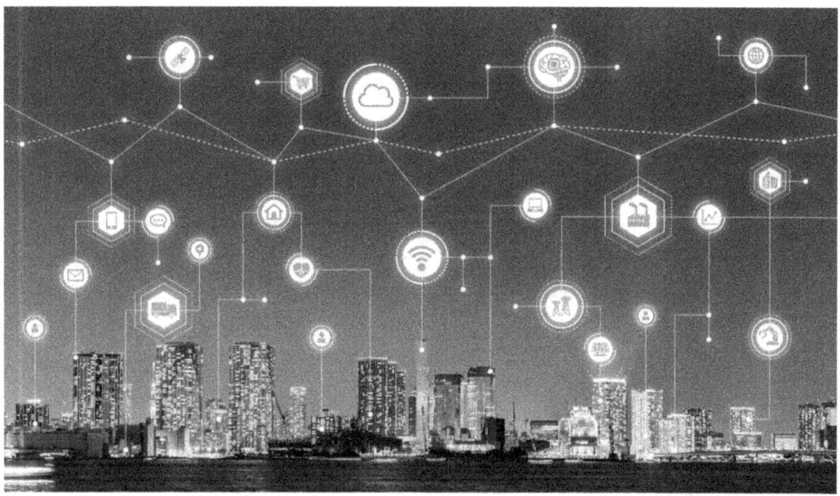

FIGURE 4.1 *A typical illustration of a smart city, showing a city center at dusk with interconnected data flows visualized in glowing blue. Image copyright iStock.*

often contrasted to orange or brown buildings or lights (Rose and Willis, 2019). Digital data is made visible as geometric patterns of blue glowing light traveling effortlessly between various digital devices in the urban environment: environmental sensors, sensors in phones, cars, street lamps, and so on. Its destination is very often pictured too: in animations, it usually flows to a cloud (literally pictured in the sky above the city) or to a set of linked boxes through which it flows; or to an operations center of some kind. The blue lines then flow back down from the cloud or out from the center to change something in a city: traffic lights or flood control systems or hazard warning signs or stock control systems. This data flow is shown as friction-free: there are no glitches, power outages, or severed cables in the smart city, apparently. The flow of data is all about ensuring that other city flows of people or traffic are equally unhindered. This blue flow is the story told about very many smart cities. This is how our urban "contemporary connectionist imaginary" (Munster, 2013: 1) is being conceived: as blue networked filaments.

The blue glow of digital networks is often superimposed onto photorealistic images of urban spaces, where they glimmer and pulse as data circulates. And like this data, the people occupying the smart city are also shown as always mobile. Pictures of flows of people constantly on the move appear in a great many visualizations of smart cities (Rose, 2018). Promotional videos about smart cities almost always begin with an aerial view of a city: the camera tracks a plane landing, or flies above a city, or zooms into the city from a god's-eye view of the globe in space. The viewer sees busy public spaces, roads full of cars and buses, crowds on train platforms and in shopping centers and markets.

This emphasis on urban crowds and crowded spaces is particularly clear in film-based promotional videos. The human in these smart city videos is represented as a mass, an "agglomeration" (Halpern, 2015: 4). Given the importance of traffic management to visions of smart cities, people are shown using many forms of transport—cars, buses, bicycles, trains, pushchairs, and, in one animation, a wheelchair—and if not driving, pedaling, or being transported, they are walking. The only time humans seem to pause in this endless movement is when they are waiting for some form of transport to arrive, which is always figured as one of the problems that a smart city will overcome (Rose, 2018).

The films and animations made by MK's local digital visualization company DVC exemplify the same qualities of flowing data, flowing people, and flowing traffic. There are many aerial views of MK in their animations. Some are maps with glowing lines representing real-time traffic density information; the aerial point of view then moves down and into fly-throughs of the city with information about parking spaces and routes appearing on screen, which in turn dissolve into the digital screen interfaces that promise to manage the flow. The visualizations frequently picture urban locations as data: a cursor hovers over a building and up pops a screen with electricity use or occupation rate. A video for the 2018 Smart City Expo uses the device of an animated map of the city to link between showcases of the work of the main stakeholders in smart MK. Several videos explaining and advertising the TSC's CAVs show similar sensations of constant and efficient mobility (Wigley and Rose, 2020). An early CGI video, *Driverless Pods* from 2014, is typical. It begins with a female voice asking us to imagine a world of clean, quiet, and comfortable transportation. The scene then opens on a set of futuristic CAV pods parked and waiting outside a building (which is in MK, although the town is never explicitly named). The CAV vehicles are shown smoothly moving without stopping other traffic. The CAV videos repeatedly demonstrate the convenience and ease of CAV use by showing them controlled by smartphone apps and the data they gather from their various sensors. Again, this seamless momentum is both enabled and shared by humans, hardware, and data.

The references to MK in these images are rather different from the work done to show Qatariness in the Msheireb Downtown project discussed in the previous chapter, however. There, we argued that the developer was keen to try to inflect a generic large-scale urban redevelopment project, designed by European architects with no knowledge of vernacular Qatari architecture, with a sense of a specific place, and that the CGIs were expected to show that place-specificity. In MK, in fact, there is also a strong sense of place which informs many of the city's smart projects. Many of the actors in MK's smart city projects have a strong sense of the history of urban innovation in MK. Some of this is about the architectural and design history of the city. Urban policy makers, politicians, academics, and planners are aware of the city's distinctive history as a testbed of innovation, and often refer to it in their talk about the city; it informs a number of council policies, including transport and economic growth (Valdez et al., 2018), and the MK2050 vision document produced in 2017; local organizations have held regular events about aspects of urban planning and design for many years in MK, to large local audiences; there is the ongoing role of MK Gallery in playing with arts-based urban experimentation; and there is the wide range of smart city projects. But there is also a conviction that MK is a relatively young city with a particular attitude to change and growth. There is a perception locally that MK is a city that looks forward to the future. According to a local app developer:

> MK has always been at the heart of our brand, since we launched, and it's really important for me. [. . .] It goes back to the attitude of the city and a lot of people talk about what the founding fathers set out to create, but, for me, the thing I always take from Milton Keynes is, actually, it's not made up of grid roads, or it's not made up of roundabouts, or landscape, or trees. For me, it was made up of an attitude of "we can do it" and we can do it our way [. . .] I always have enormous respect for that ethos, if you like, and that's what I take from how Milton Keynes was built and how I would really like to see it continue to grow in the future.

This is a feeling that defines MK as a particular kind of place oriented to the future. It is also a sense of place that aligns with smart experimentation. The city's director of Strategy says that "it's a place where a lot of people use the term a 'living lab' but, actually, I think it's got that attitude of it can be."

However, while much of smart activity in MK is driven in part at least by a distinctive sense of place, in the marketing of the city as smart to the external world, the place-specificity of smart innovation in MK has to be presented carefully. While the place itself must be visible to potential investors as a useful urban-lab site, the city cannot be so specific that what is experimented on in MK cannot be exported to other places. After all, the point of an urban lab is to test a technology that can then be "scaled-up" and sold in other places as well. In the various visualizations of MK as a smart city, therefore, there is a different sense of place being conjured than that in Doha CGIs. Many of MK's smart city visuals use recognizable MK locations but do not name them as such (Wigley and Rose, 2020). CAVs in MK, for example, are focused on the "last mile" problem: getting people from a large transport hub in the city center to their specific destination. The most significant space in the promotional videos about autonomous vehicle work in MK is then not MK itself but rather "the last mile." A 2014 animation by DVC ends by emphasizing how innovation trialed in MK will benefit "other cities right around the world" and the 3D model of MK that accompanies this claim is of white and gray building blocks, like an architect's

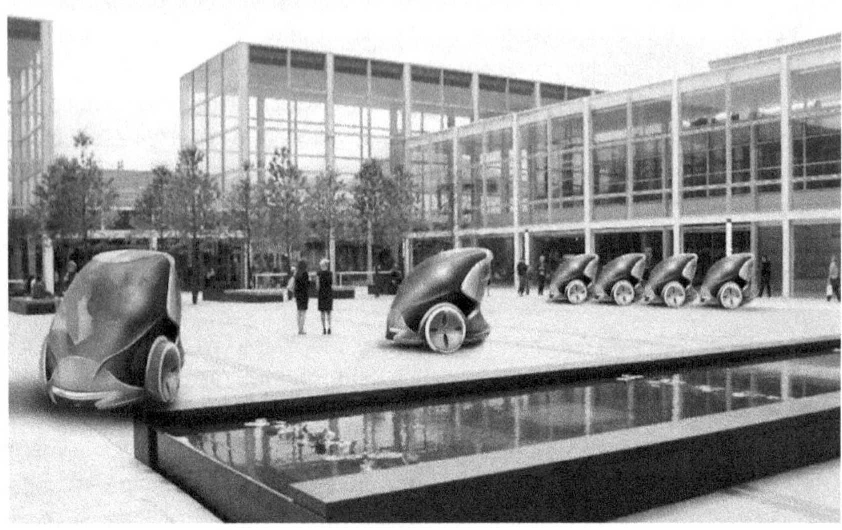

FIGURE 4.2 *Driverless connected and autonomous vehicles visualized in MK's shopping center in 2015. Image copyright the Transport Systems Catapult.*

maquette. The city is visible as a backdrop but absent as a unique place. The background of Figure 4.2, for example, is actually MK's central shopping center, but works in this image as an unnamed generic urban place.

These images are not seductive pictures of lovely, glamorous urban locations then. What they show instead is a city—any city—that flows. Data, bodies, and cars are all shown as constantly mobile in an urban location. Bodies move, data moves, and data about bodies moves as well as all sorts of other data. This movement is visualized in order to demonstrate that its extraction can enable better urban management based on better knowledge about a city. But another movement is also implied, namely that of the technology of data extraction and analysis which can be implemented in another location. So the image has to represent both a particular city, and any city. The conceived space being pictured in these smart city visuals is not only MK as a specific urban location but also something more abstract: a city that can be smartened by having the sparkling blue net of data laid over it.

4 Perceiving the New Urban Aesthetic in a Smart City: The Flow of Bodies and/as Data

An aesthetic of flow pervades conceived images of smart. The smart city is imagined as a network of flowing data overlaid on the material urban environment, pulsing through it, that data generated by all kinds of sensors, including smartphones, and collated and displayed on many screens. Just as data moves, so must everything else: traffic, water, people. These conceived representations are an interpretive grid that shows and values a very particular urban experience: one that is constantly on the move. This section now moves to explore the perceived aspect of this textured smart urban space.

The aesthetic of flow is one part of the wider reconfiguration of embodiment in what Deleuze (1992) called the "society of control." In his short essay of the same name, Deleuze suggested that the society of control was succeeding the biopolitical and the governmental regimes of organization described by Foucault. He emphasized the centrality of digital code to societies of control, and suggested that code was shaping movement: where people could go, what information they could access, what was open to some bodies and closed to others. The body in a society of control is "undulatory, in orbit, in a continuous network" (Deleuze, 1992: 6), just like smart city embodiment. Mobility depends on "the computer that tracks each person's position" (Deleuze, 1992: 7); one day, your privileged thumbprint may no longer open your CAV door, or your banking app may not recognize your face. Social power no longer corresponds to specific bodies, he suggested, but rather into software, and what software permits.

There are many issues with Deleuze's account, written in an early moment of the digital mediation of cities. For example, it is not the case that societies of control are simply replacing earlier forms of power. Smart cities contain many kinds of power (Isin and Ruppert, 2015: 37). As noted in Section 2 of this chapter, many critics have argued that the inhabitant of the smart city can be understood as a neoliberal individual produced by governmentality (Vanolo, 2014). Most importantly perhaps, software continues to differentiate bodies according to class, race, gender, sexuality, and many other forms of social differences that have historically been understood as grounded in corporeality rather than data that can be generated about that embodiment (Benjamin, 2019; Browne, 2015).

Nonetheless, Deleuze's suggestion that other kinds of power dynamics are emerging in digitally mediated cities aligns with this book's account of the emergence of a new urban aesthetic, and it has been taken up in some discussions of smart cities. Krivý (2016), for example, begins with Deleuze in his account of smart cities, and then moves on to second-generation cybernetic theory to explore how smart cities aim to use data. In his account, humans in smart and platform cities are just one among many sensors embedded in data feedback circuits (see also Thrift, 2014). Humans are converted into data via, for example, heat cameras, or the swipe of a transport system card, what they say as they tweet, or, most often, the data generated by their use of a smartphone and its apps. So too are many other elements in the urban environment: particulate matter in the air, rivers and drainage channels, the location of buses and the number of cars passing through a junction, the number of smartcards being swiped at a metro station. Managing cities by working with this sort of data leads to a form of urban management that is about real-time adaptation and optimization, says Krivý (2016; see also Luque-Ayala and Marvin, 2020). This is precisely what many images of smart cities exemplify, as the previous section noted: bodies among other objects generate data, and that data is used immediately to manage cities in the most efficient way.

In this account, data-driven urban management depends on both human and algorithmic analysis. Data is "fed back through big data sorting that is designed to collate seemingly unrelated sets with the intention of producing novel relations" (Clough, 2018: 107). There are consequences for how embodiment is thought here. Rather like the body in Deleuze's account whose ability to act is not a constant but depends on software processes, the bodies and sensoria that emerge in this cybernetic notion of digitally mediated urbanism are not stable. They emerge as a consequence of the patterns identified by the algorithms at work on the data generated by—for example—smartphone apps. As Cheney-Lippold puts it:

> algorithms make our data speak as if we were a man, woman, Santa Claus, citizen, Asian, and/or wealthy . . . these algorithmically produced categories replace the politicized language of race, gender, and class with a proprietary vocabulary that speaks *for* us—to marketers, political campaigns, government dragnets, and others—whether we know about it, like it, or not. The knowledge that shapes both the world and ourselves online is increasingly being built by algorithms, data, and the logics therein. (Cheney-Lippold, 2017: xiii)

This returns us to the concerns about privacy and surveillance in smart and platform cities. But it also relates to the aesthetics of inhabiting the perceived space generated by such data processing. This is touched on in Gabrys's (2014) discussion of smart citizenship as it is articulated in a range of smart sustainability projects. She suggests that the production of digital data produces smart citizens as subjects whose emergence is contingent on the data generated about actions and events and its processing through the technologies and practices of computational urbanism. She describes the people in this process as "ambient and malleable urban operators that are expressions of computer environments" (Gabrys, 2014: 42–3). They emerge from the measurement and calculation of the materiality of urban spaces, and from the processes of data circulation. In that sense, they inhabit perceived urban space. Generated by the technological measurement of urban life, these embodiments are part of "the more material, sensuous dimensions of the media; the very stuff in terms of tools and infrastructures for mediation that make up our everyday environments" (Jansson, 2013: 282).

Our interest in these arguments is about what being an "ambient and malleable urban operator" in perceived urban spaces might feel like. There are some more theoretical accounts of that feeling. Thrift, for example, drawing on a language of affect, suggests it might be felt as a certain form of "push" (Thrift, 2014: 9) or perhaps it is something like "a direction that might be traced or felt" (Munster, 2013: 39). The next section turns to a suite of smart city smartphone apps in MK to explore the perceived texturing of this version of the new urban aesthetic as a sensation of flow.

5 Appified Sensations in Smart MK

This section brings the previous discussion on a perceived new urban aesthetic to bear on a group of smartphone applications, designed as part of a number of smart city projects in MK between 2014 and 2018 (see also Rose et al., 2021). Smartphone apps have not received much attention in the literatures on smart cities. However, apps' integration into various platforms and databases is an important means by which data in a smart city can be generated by both companies and city authorities (Barns, 2018; Thatcher, 2017). Apps are created as part of many smart city projects. They are an interface between urban citizens and smart data infrastructures, and between people and digitally mediated urbanisms more generally (Graham et al., 2013; Leszczynski, 2016; Schwanen, 2015; Thatcher, 2017; White, 2016). As the chapter's introduction noted, apps are also a key device in the digital mediation of bodies.

This discussion of apps is based on interviews with the people who commissioned, designed, and built eight smartphone apps in MK. Like the previous chapter then, this chapter engages with the aesthetic labor of design professionals, and pays attention to what they say about what their labor was intended to achieve. The eight smart city apps are Motion Map, Breastfeeding Hub, Redway Reporting, Garden Monitor, Eyeware, Discovering MK, MK:3D, and Age Friendly Map. The first four were facilitated through the MK:Smart project. Motion Map was part of MK:Smart from its inception and aimed to support and influence how people moved around the city (Valdez et al., 2018). By offering real-time information on transport and travel options and their current usage, app users traveling around the city could choose the most convenient but also the least carbon-intensive mode of transport. The Redway Reporting app and the Breastfeeding Hub both emerged from the Citizens Work Package of MK:Smart, which facilitated local digital innovation. The former allows problems with the city's 200 kilometers of cycle and walking pathways to be reported to the council. The latter carries information about breastfeeding and also allows users to rate locations in the city according to how welcoming they are to breastfeeding. The Garden Monitor app was a side-project for MK:Smart's lead data scientist, who is a keen gardener, and was intended to optimize garden watering using data on weather and soil conditions. The Age Friendly Map app was a collaboration between MK Gallery, Age UK Milton Keynes, and local digital design company DVC, and aimed to make the city center more accessible to older people by giving them information about the city center that was relevant to their needs. The Royal National Institute for the Blind collaborated with the Transport Systems Catapult to produce the Eyeware app, which allows users to experience different forms of visual impairment. Discovering MK was a collaboration between the council and the city's Arts and Heritage Alliance; it is a geocached walking tour around some of the city's historical landmarks which displays information about the landmarks when the app is nearby. The MK:3D map was a collaboration between DVC and Milton Keynes Council; using touch gestures, it allows users to move over a 3D model of the city and pause over locations to get information about the location and its employment and investment opportunities.

Only the Breastfeeding Hub and Eyeware remain in the app stores at the time of writing; the others were either prototypes or have been taken down. Only Discovering MK and the Breastfeeding Hub appear to have been well-used locally (the latter now has clusters of users beyond Milton Keynes). In their range of partners, funders, and success (or failure), these apps are typical of the messy variety of actually existing, urban-lab type smart cities like MK (Cowley and Caprotti, 2019). Their data came from different sources, took different forms, was shared in different ways, and had somewhat different anticipated effects. It is important to note that only one—the Motion Map—was designed to work with large-scale, real-time datasets. Most of these apps were about much smaller, local data circulations: the app itself provided data (in the form of information about the city for the user), and might ask its user to upload information in return (to rate a location or upload a photograph or report a problem). The discussions explored in the previous section mostly assume large datasets algorithmically analyzed, and these apps do not fit that model. Nonetheless, as we will see, the apps' design presumes and enacts an aesthetic of flow. Those assumptions displaced other forms of differentiating between bodies.

The apps' makers all assume that sharing digital data will improve users' experience of living in the city: hence, data must flow. They also all assume the app users' mobility: bodies must move. Five of these apps aim to facilitate bodily mobility: the Motion Map and the Redway Reporting app are intended to improve travel experiences (the prototype of the Motion Map even proposed

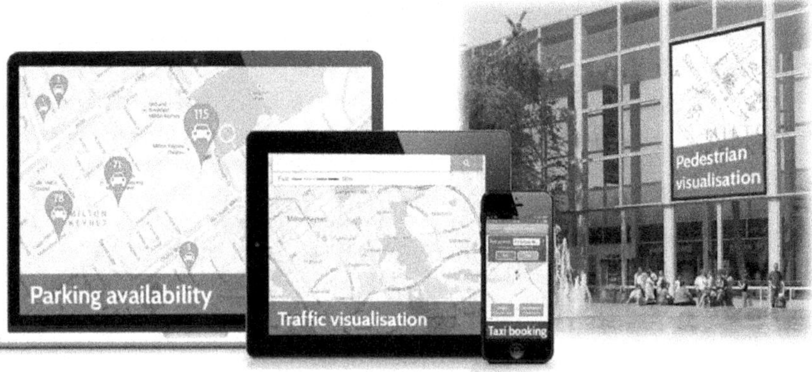

FIGURE 4.3 *A publicity image for the Smart:MK Motion Map project. Image copyright the University of Cambridge.*

using heat sensors to share real-time information about where MK's shopping center was busiest, so visitors could avoid the crowds: see Figure 4.3); the Breastfeeding Hub wants to make the city more accessible for breastfeeding and the Age Friendly app aims to make the city more accessible to older people; and the Discovering MK reveals its geocached information about a heritage site only when the app has been taken to the site. The MK:3D app allows the viewing body to zoom and fly over the city as the screen is swiped, enacting the untethered gaze of much digital visuality (Rose, 2018). Eyeware, meanwhile, is an augmented reality app that allows its users to imagine how difficult moving around would be with a visual impairment. Even the Garden Monitor aims to tell gardening bodies when it is most sustainable for them to go to their garden to water it. That is, each app assumes mobile usages.

The app designers gave lots of reasons for this emphasis on mobility. The local transport infrastructure, and especially its fragmented and unreliable public transport system, was a motivation behind Motion Map. Council officers and voluntary sector workers often mentioned that getting people more mobile was a means for improving public health. This was the case for Discovering MK and also the Breastfeeding Hub; if women felt more comfortable breastfeeding in public, its advocate reasoned, then more women would do it. Another reason offered by the designers of the Discovering MK app was that the app enabled people to find out more about the city: "the main intention, obviously, was to get people learning about the arts and the culture in Milton Keynes and visiting those particular spots and locations, and that was the primary function." Yet another reason given for app development was environmental sustainability: the Garden Monitor app was introduced by its designer with the factoid that at current levels of population growth MK would run out of water by 2030, and the Motion Map was also about encouraging more sustainable as well as more efficient modes of transport (Valdez et al., 2018).

Mobility is central to many apps, which are designed to build on the affordances of the smartphone as a portable device and the specific notion of "location" built into their software and hardware (Thatcher, 2017). However, these apps also assume not only mobility but also a specific kind of responsive user. As well as their design as part of a smart city and smartphones, these apps assume that mobility will be generated by providing data about locations. If sufficient data of the right kind is provided by the app, mobility will result. A good example of this logic is the Breastfeeding Hub app (see Figure 4.4). Breastfeeding has been a contested practice for many years in Western cultures, where breastfeeding has been expected to be done in private. More recently some public space for this practice has been reclaimed. The Breastfeeding Hub assumes that this success can be amplified by giving more information about where it feels comfortable to breastfeed in MK in public. It does this by sharing how breastfeeders have rated specific locations in the city. Thus, the assumption is that the right kind of information will enable more access to more places. The Discovering MK app also assumed that enabling people to unlock information about specific sites in the city only when they were at that site would incentivize bodies to go there and then with what they learn, to engage in the city in the future. This emphasis on accessing information as a means of generating specific behaviors assumes that predictable, desired behaviors will be generated by specific information (and see Thatcher, 2017). Similar assumptions are also made in much work on transport choices, and underpinned discussions about the Motion Map app. Mobility is assumed to happen when appropriate information is provided.

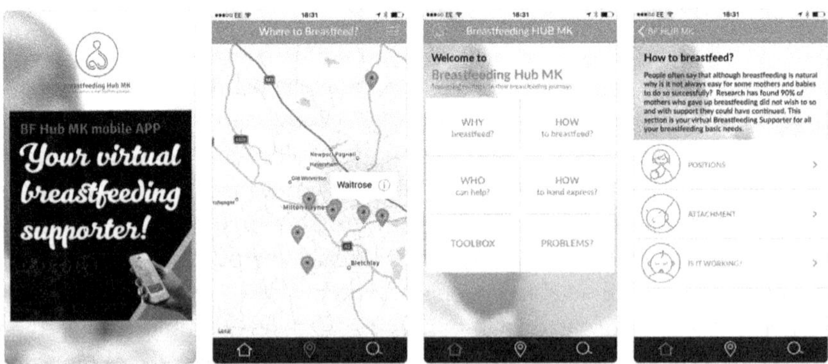

FIGURE 4.4 *The Breastfeeding Hub MK smartphone app. Image copyright the Breastfeeding Hub Ltd, photo credit David Isaacson.*

This assumption that data leads to specific urban process—in this case, mobile bodies—constitutes perceived urban space. The data, it is assumed, is encountered directly and has effects. There is no process of interpreting or questioning the data (Rouvroy, 2013). Instead, the app user becomes "an operational component of the infrastructure" both digital and expressive (Luque-Ayala and Marvin, 2020: 175).

What followed from this model of behavior was a widespread assumption among the app designers in MK that simply using the app made other social differences irrelevant. The designer of the breastfeeding app insisted that an app gave equal access to information and support:

> I deal with teenage mums. I deal with mums who maybe have addictions or whatever, you know, all sorts of social issues and challenges. And they struggle, but nearly every single one of them will have a smartphone.

In relation to the local heritage app, its council lead remarked that "it doesn't seem to matter who we aimed it at. There's an awful lot of elderly people on their own out doing it, walking and doing it and there are an awful lot of families," and its designer elaborated, "it's really nice to see that we've got so many different types, and so many different individuals, and families, and age groups, and ethnicities, and everything, that have downloaded and used the app." It is useful to recall Cheney-Lippold's comment here that "algorithmically produced categories replace the politicized language of race, gender, and class" (2017: xiii). While not produced by algorithms of the sort referenced by Cheney-Lippold, the mobilized body nonetheless also seems to displace many other possible ways of differentiating among appified bodies. Instead, the only thing that mattered was the mobility of the body and what was assumed to follow from that: better health or less environmental impact. Across these apps, bodies were reconfigured as mobile by the generation, extraction, and provision of data about the city.

This section has discussed a group of smartphone apps designed as part of a smart city. The smart city is an ordinary one—MK—and the apps might be described as rather ordinary too. Only one had ambitions to be a city-wide urban data platform (Motion Map) and, as it turned out, its functionality was overtaken by the arrival of Google Maps (Valdez et al., 2018). Nonetheless, in their assumptions about their users, these apps reiterate the new urban aesthetic of flow, this time through the perceived aspects of its triadic space. The apps produce a sensory orientation to movement, which is valorized as a good thing because, for example, it is about bodily health or urban citizenship or environmental sustainability. The apps demonstrate the emergence of perceived urban space as a consequence of (what is assumed to be) an unmediated response to a particular kind of digital material infrastructure: the networked circulation of data and its automatic processing and analysis. The aesthetic labor designing these MK apps' interfaces and their data circulations parallels this infrastructure in its assumptions that mobility is desirable and can be induced simply by access to the correct information. This is the push or direction of this expressive infrastructure of flow (Munster, 2013: 39; Thrift, 2014: 9). And while this direction is not explicitly gendered, classed, racialized, or sexualized, it does position bodies who do not move as a problem: a point we will return to at more length in Chapter 6.

6 Perceiving the New Urban Aesthetic of Flow, Again

This section circles briefly back to the smart city imagery discussed in the second section of this chapter. The second section described that imagery as conceived. It is an abstracted representation of an imagined kind of city. Smart cities like MK are shown in very particular ways: as saturated by networks of flowing data, flowing people, flowing light. This conflation of bodies and data by visualizing flowing streams of light is particularly clear in a scene that is used in very many smart city visuals, of highway traffic at dusk or at night, only the headlights of moving vehicles visible, the moving stream of lights blurring into the animated lines of light which are the conventional sign of data streaming through Wi-Fi networks. Cars, bodies, and data become melded into one circulating flow. This section suggests that we should also consider these images as constituting a perceived urban space. Just as automatic human responses to data are presumed to enact perceived urban space, so mobile data also intervenes somewhat directly in the creation of these images to render them as perceived as well as conceived.

Section 2 discussed how smart city images are digitally produced to visually represent urban space. Sometimes they mimic older and well-established forms of visual imagery: a photograph, say, of a smart operations control center or of an urban sensor, or a video of a person talking or of urban crowds. Often, like the Doha CGIs, they claim to reference a desired future, which will be delivered by existing technology. As the conclusion to Chapter 3 noted, this sort of conceived image can be understood as a "solid representation of a solid world based on the sound principle of geometric projection" (Hoelzl and Marie, 2015: 7). But in terms of its own digital materiality, smart city imagery is not solid at all. It is remarkably malleable, in fact. It not only shows a city in constant motion, the image itself is constantly moving and morphing. Animated images are in constant motion with endlessly changing points of view.

Digital visualizing software is an important enabler of this visual mobility. As the previous chapter noted, Hoelzl and Marie (2015) describe digital images as softimages, because they exist entirely through software and are thus infinitely adaptable. A digital image only becomes an image as part of complex interconnections between hardwares and softwares. In her discussion of "the importance of media-specific analysis," Hayles notes that a digital image is

> produced by layers of code precisely correlated through correspondence rules, from the electronic polarities that correlate with the bit stream to the bits that correlate with binary numbers, to the numbers that correlate with higher-level statements, such as commands, and so on. Even when [images] simulate the appearance of durably inscribed marks, they are transitory images that need to be constantly refreshed by the scanning electron beam that forms an image on the screen to give the illusion of stable endurance through time. (Hayles, 2004: 74)

Many smart city images display this transitory aspect of their own digitality. They are visibly edited in all sorts of ways: cropped, color and focus intensified, and other visual elements added (like blue data networks). Fantastical images of smart futures are invented which bear no relation to any preexisting city. Digital images are also designed to be shared, between devices and between devices and platforms (Rubinstein and Sluis, 2008) and to be configured by algorithms (MacKenzie and Munster, 2019): to be sorted by date, to capture smiles, to tag faces, to be grouped by location, to be searched and retrieved (Thylstrup and Teilmann, 2017). Hence they circulate widely on social media, and through Google Images and stock image platforms (Frosh, 2003). There are many ways in which softimages demonstrate their merger with software.

This point was made in the previous chapter. In the case of the Msheireb Downtown CGIs, however, a lot of work went into stabilizing their meaning and the conceived, representational aspects of what they showed. Here, however, in the new urban aesthetic of flow, we want to emphasize that digital technologies constitute images which are inherently and also visibly perceived as animated—indeed, even the Doha images continued to change as Msheireb Downtown's designs changed. "Animation" is a term discussed at length by Levitt (2018), who claims that animation is "the dominant medium of our time" because it moves us from "questions about ontology, category, and being to ones of appearance, metamorphosis, and affect" (Levitt, 2018: 1, 2). Animation's logic is not based on reference to a real; instead, animations envision metamorphosis, erasure, and resurrection. Levitt is clear that animation is not dependent on digital visualizing technologies. Animation has a long analog history. Nonetheless, digital technologies are especially suited to generate transformative novel sensations. Digital images can show what is not there, and they constantly change and morph. Transformation is at their core—even in the ways they are perceived.

This description of animation can be applied to very many smart city images. They not only show things on the move, they are constantly on the move themselves. Many are animations: YouTube and Twitter videos, graphics, and cartoons. They constantly shift and change form. Content mutates, and different visual formats flow and merge one into another. A map morphs into a data network, an aerial photograph of a city transforms into massing study, a city is converted into a set of icons, a building becomes a data display, to mention some of the things DVC does in its MK videos. And they circulate around many different displays, as already noted.

Hence in these images of smart, data flow is not simply pictured as blue lines pulsing across city street. Data flow is the organizing principle of these images. As a way of seeing, this has implications for the viewer. Levitt suggests that animations tend to merge and exchange image space and body space (Levitt, 2018: 83). No longer a single point of view framed by perspectival techniques, the spectator becomes positioned as a constantly mobile point of view, zooming and hovering and sweeping through an environment that seems to have no frame. Elsaesser (2013: 240) describes this unanchored viewing, tracking seamlessly through spaces from the nano to the planetary, as "the default value of digital vision" (and points to its nondigital precedents in a range of efforts to create convincing three-dimensional films). In a similar way, smart cities are now insistently visualized through multiple interfaces, in different formats, genres, and media: there is no single frame, no nest of scales, no coherent territory. Different images merge and blend, and the same image reappears in different contexts. References to a real become references to multiple reals become a seamless dissolution of one real into another becomes more visual flow, exemplified in videos evoking the smart city which seamlessly mutate between multiple types of visualizations of city spaces (Rose, 2018). Like the anticipated inhabitants of the smart city then, moving in concert with data flows, the viewer of smart city visualizations is also invited to become a moving part, positioned in relation to the image not just by its representational references but also by its animated flow. Like the data it pictures and the data that generates it, the smart city image reconfigures its viewer as flow.

7 Conclusion: Texturizing Flow

This chapter began by noting that claiming to be a smart city is part of the branding of many cities. Smart cities are not simply about technologies, infrastructure, devices, and data. They are also caught up in conceived and perceived urban spaces. They have stories told about them, and in that sense, smart cities are constituted through conceived forms and representations of urban space. The chapter illustrated the particular kind of place representation entailed in branding a city as smart. While many city authorities aim to promote their city as a smart city, this smartness has to appear both to represent that particular city and to suggest that its smartness can work elsewhere. This paradox was explored in the chapter's case study city, Milton Keynes. Local smart actors are keen to express what they feel is the uniquely innovative, action-based and future-oriented feel of MK as a place. But the branding of MK externally cannot be about the city's uniqueness because the point of many of its smart projects is that they are trialed in MK and then deployed elsewhere. Smart has to look as if it can function anywhere.

Picturing the flow of data manages this paradox because while the data itself is assumed to represent a particular city, a data infrastructure and network can be overlaid onto any city. The chapter went on to identify an aesthetic of flow as typical of smart city promotional imagery. What that flow shows are things on the move, including data, and bodies, and the data generated by bodies. This flow is conceived. It both represents a particular kind of city and valorizes that smart city as the best kind of urban future. The chapter's discussion of a group of smartphone apps designed as part of MK smart city projects suggested that this presumption of the mobility of data and bodies was not only a conceived aspect of this flowing aesthetic, however. It is also perceived. Mediated by a smartphone app, the movement of bodies is assumed to be desirable, smooth, and straightforward. It is assumed that if the correct data can be harvested, analyzed, and offered, human interpretation of that data is unnecessary. The users of apps—and, we might speculate, of smart cities more generally—are understood as simply responding to the data they are given in the appropriate way. In unpacking the material, sensuous stuff of app smart tools, the chapter identified the emergence of perceived urban space.

The sensibility which emerges from the algorithmic analysis of data is not one that reflects or interprets. It is co-constituted with the material flows of data and its operations: data which induces action, movement, rather than interpretation. Section 5 of this chapter showed that even when big data is not actually present, this embodiment emerges in the design of smart city smartphone apps. Everybody who uses a smartphone is categorized just through that action, as a "user," who will respond directly to the provision of appropriate information/data by becoming mobile. Other means of categorizing bodies, such as the signs of class, race, gender, and so on are displaced. They do not entirely disappear, as Chapter 6 will discuss. But the perceived aspect of the flowing new urban aesthetic is a powerful emergent presence, and the final section of this chapter tracked its reconfiguring of the viewer of smart urban branding images.

This chapter has conceptualized the new urban aesthetic of flow as both conceived and perceived, then. Its texture is both representational and sensory. It is one way in which "incorporeal flows of information and combine, in convergent and divergent ways, the capacities and functions of carbon materialities with those of information flows" (Munster, 2006: 23). Of course, this new urban aesthetic does not mediate the entire experiencing of any city. Different apps, different data, different practices, different platforms all intersect in what Barns (2020) describes as a complex urban platform ecosystem. These will offer other sensational seductions and also glitches (Leszczynski, 2019a, 2019b). There are other kinds of mobility, of course, with different sorts of effects. And there are other digital mediations of embodiment. The bodies that appear in the CGIs of Msheireb Downtown are also bodies converted into data, for example, in this case into photographic JPEG or RAW files, to be cut and pasted from one database to another, one scene to another, rendered brighter or translucent (see Figure 3.2) through digital tools. Those digital manipulations of how bodies look in scenes that are meant to look realistic enough to be seductive is another example of how urban sensations are becoming variously mediated in new ways by digital mediations, so that what is visible is shaped by the adaptability, mobility, and searchability of digital images (Rose, 2021). That does not leave the corporeal body as an unmediated excess to this digitality, however. As this chapter has demonstrated, embodied sensation can also be deeply entrained in both the perceived and conceived spaces of the new urban aesthetic of flow.

5

The Lived Aesthetics of Urban Social Media

Anticipating the Culture Mile's Future

1 Introduction

In recent years, the sharing of images of urban environments on social media has become pervasive, radically transforming the ways we represent and experience urban places. Social media platforms like Facebook, WhatsApp, Uber, Instagram, FourSquare, Deliveroo, Lyft, and many others now mediate many everyday embodied experiences of urban spaces. They are "rapidly changing how people experience cities, and even how cities work" (Rosenblatt quoted in Leszczynski, 2020: 190). While urban scholarship has paid some attention to how social media platforms generate big data, and to their social networking capacities—to organize urban campaigns and protests, for example—there has been relatively little attention paid to the intersection of social media with one of the major themes of this book, urban branding. Yet architects, developers, city councilors, tourism agencies, and retailers are increasingly aware of the branding power of social media, and of Instagram in particular: "[f]or a place to be shared on Instagram is no longer a chance by-product of a photogenic design, but a primary concern that drives the ambitions of clients and designers. The idea of 'doing it for the "gram"' has moved from the preserve of Like-hungry teens to board meeting discussions and multimillion pound budgets" (Wainwright, 2018). Instagram had more than a billion monthly users in 2020 (which is 12 percent of the world population), and has "grown to become one of the most influential forces in the way our environments are being shaped" (Wainwright, 2018).

Boy and Uitermark (2017, 2020) have provided probably the most in-depth analysis to date on how a city, in this case Amsterdam, has been reshaped and reassembled on and through Instagram posts. Their emphasis is on how

"Instagram users act out aesthetic and lifestyle ideals as they craft images and strategically display aspects of their life worlds" (2017: 622) and thereby reinforce processes of gentrification by elevating and featuring some places center stage "while other remain peripheral or are altogether ignored" (2017: 622). While we fully agree with their findings, we expand in this chapter one aspect of their discussion mentioned but not explicitly explored, namely how a digital platform such as Instagram mediates lived urban experience. In particular we would like to pick up from and develop their concluding statement that "[m]ateriality or visceral experience do not become less important but are increasingly intertwined with images and messages circulating through a range of communication circuits" (Boy and Uitermark, 2017: 623).

Manovich (2017) clarifies how Instagram needs to be understood as part of a broader expansion in mobile photography culture, yet with the difference that "in contrast to earlier photo services such as Flickr, 'Instagram was meant to be an app for sharing pictures with people, not an app for photographers' (Sandhya Ramesh on Quora 4/23/2015)" (Manovich, 2017: 41). He further points out that Instagram's visual aesthetic is a crucial feature of its popularity and global appeal. Aesthetics matter greatly because as in any style of photography, Instagram is "about making visual images that communicate through their techniques, styles, and visual choices—and not only content" (Manovich, 2017: 40). Furthermore, because of the app's large range of filters, it is easy for Instagram users "to use [the app] in aesthetically sophisticated and nuanced ways . . . [and] to take any photo and make it visually interesting and appealing" (Manovich, 2017: 40–1). He concludes that "if Google is an information retrieval service, Twitter is for news and links exchange, Facebook is for social communication, and Flickr is for image archiving, Instagram is for aesthetic visual communication" (Manovich, 2017: 41).

Hence, having explored CGIs as exemplary visualizations of planned urban futures in Chapter 3 and app-driven circulations as enacting new forms of digital urban embodiment in Chapter 4, this chapter focuses on Instagram to further deepen our discussion of the new urban aesthetic. We examine how a particular area in London is pictured on Instagram. Chapter 3 discussed CGIs in terms of the *conceived* aspects of the urban space they mediated, examining how urban space was represented. Chapter 4 paid attention to the role of smartphone apps in new experiences of *perceived* urban space. This chapter emphasizes the *lived* aspects of a new urban aesthetic's spatiality. Lefebvre referred to lived urban space as the consistent and habitual enactions of urban life. Lived urban space saturates everyday life and holds it together as a normal way of doing things (Jansson, 2013).

The case study is a cultural regeneration project in the City of London called the Culture Mile, and for the purpose of this chapter we focus on one of its landmark buildings and immediately surrounding area: Smithfield Market, currently still functioning partly as a wholesale meat market and the key focus of redevelopment in the area. As part of a "cultural corridor," the west end of Smithfield Market is earmarked to be converted into the new Museum of London by 2024. The chapter looks at the aesthetic labor of developing a social media branding strategy for this urban redevelopment scheme. But we also look at the sense of place and uses of Instagram in the Culture Mile area by its workers, visitors, and residents since, as Chapter 2 noted, aesthetic labor is by no means only the preserve of design professionals. Examining the lived aspects of the Smithfield Market area on Instagram enables us to identify the emergence of a third version of the new urban aesthetic: a *dramatic* aesthetic which is shared by very many Instagram users. As we will show, the increased mediatization of urban environment via social media leads to a particularly staged organization of the urban environment, in which bodies experience intense feelings and emotions for a limited time and which are photographed, posted, and shared across the world. We also further the book's argument by suggesting that the conventions of Instagramming mean that the branding and the non-branding versions of the Culture Mile, as they are staged on Instagram, are difficult to distinguish from each other.

We are particularly attentive to how this dramatic aesthetic is structured through a complex temporality which both intervenes in the present and anticipates a future, and in Chapter 6 we draw on Rancière to examine the power relations enacted in this particular temporal distribution of the sensible. This chapter begins, though, with a description of the case study and how the Culture Mile's branding can be understood as an example of anticipatory urbanism. This is followed by a discussion of how anticipatory urbanism and urban branding have increasingly become mediated by social media, and the Culture Mile's social media branding strategy. This allows us to define the dramatic urban aesthetic that the Culture Mile project is aiming to enact. The last part of the chapter examines the ways the area is Instagrammed using #SmithfieldMarket and #CultureMile, and this allows us to identify how that dramatic new urban aesthetic is in fact widely shared across Instagrammatic mediations of the area. We conclude the chapter by reflecting on how digital platforms such as Instagram are recalibrating the aesthetic staging and everyday experience of this area.

2 The Case Study: The Culture Mile, Smithfield Market, and Anticipatory Urbanism

The Culture Mile area of London encompasses the north-west corner of London's historic city from Farringdon to Moorgate. It covers 45 hectares: 15 percent of the total area of the Square Mile (see Figure 5.1). It was

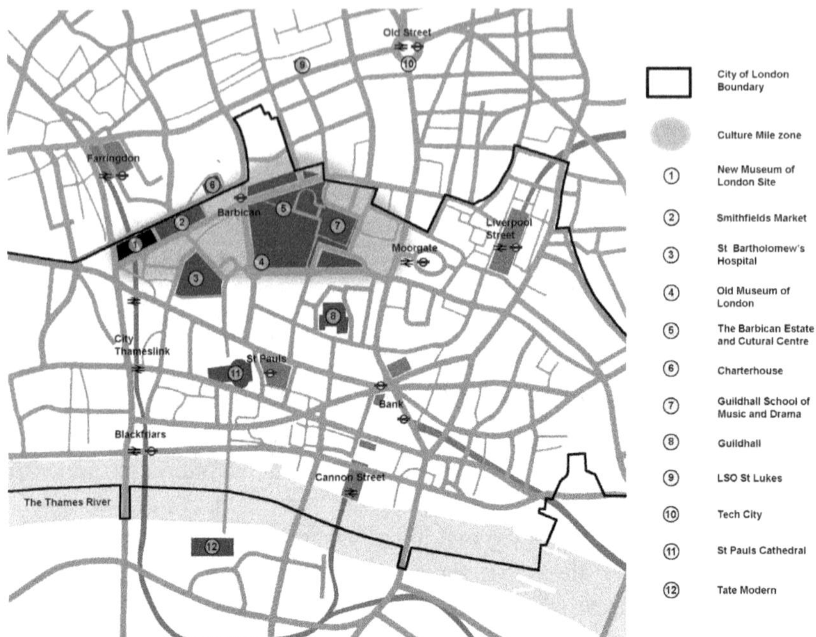

FIGURE 5.1 *The Culture Mile. Image copyright Mónica Montserrat Degen, designed by Isobel Ward.*

identified as an "Area for Intensification" by its local authority in the City of London Corporation's Local Plan 2015–20. It is also part of one of Europe's largest future cultural capital projects, unveiled in 2017 as the Culture Mile.[1] As a "designated area for development," the Culture Mile area, which encompasses the Smithfield Market area and the Barbican Estate, will be extensively redeveloped over the next decade. The transformation is driven by a public-private partnership organization supported by the Corporation of London which coordinates a number of partner organizations including the Barbican Arts Centre, the London Symphony Orchestra, Guildhall School of Music, and the Museum of London. The various cultural institutions take different responsibilities for the Culture Mile implementation under a common place brand, claiming that the "Culture Mile has the *potential* to become a global hub for creativity, innovation and learning, bringing together culture and commerce to develop the platforms, infrastructure, collaboration, skills and support to enable creativity to flourish and deliver economic growth and social mobility for London" (Culture Mile London, 2020, our italics). The expectation

[1]www.culturemile.london

is that "the latent *potential* within the area's creative sector is unlocked; adding over £4 billion per annum to the output of the City of London and generating up to 50,000 new jobs" (BOP Consulting and Publica, 2019, our italics). Note the use of "potential" in line with Anderson's (2010) anticipatory urbanism as we discuss at the end of this section.

Much of the area covered by the Culture Mile, such as parts of the street layouts, have remained unaltered since the 1870s and follow mainly a medieval pattern. The area has a rich mix of architectural styles from medieval buildings such as the eleventh-century St Bartholomew the Great Church and the fourteenth-century Charterhouse monastery and almhouse (still in use), to the Barbican, the famous 1960s Brutalist housing estate and arts center which stands next to glittering facades of austere twenty-first-century high offices—to Smithfield Market, Britain's oldest and largest wholesale meat market housed in a grand Victorian building. As one Culture Mile marketing manager explained, "it is fascinating in terms of the layers, the different eras, you can still see within one space." However, as the area is close to the financial center of London, it has not been an obvious destination for London's tourists. The Culture Mile project plans to change this. A central part of its efforts entail the Museum of London moving from its current location in the Barbican and into the vacant market buildings of West Smithfield Market in 2024, a move calculated to increase visitor numbers to the Museum from 900,000 to 2 million a year (Museum of London Strategic Plan 2018–23, 2018). The London Symphony Orchestra is planned to relocate to a new building on the Museum of London's current site, and to support access to these venues and promote the area where "creativity is fast becoming the most valuable currency" (Barbican 2019), two Crossrail stations are expected to open in Farringdon and Moorgate in 2021.

The beginnings of the Culture Mile can be traced back to 2010 when the City of London's first Cultural Strategy (2010–14) was approved. The Corporation of London, the local municipal governing body, established the Cultural Hub Working Party in the same year, which stated that its "vision for 2017 is to see the City's identity as a cultural hub strengthened in its own right, alongside its status as a financial centre" (City of London Corporation Cultural Strategy 2012–17 quoted in Cultural Hub Working Party Report to Policy and Resources Committee, 2013). The Corporation is in fact the fourth largest cultural funder in the UK due to hosting in its borough many cultural institutions, yet it is not often acknowledged as such, and the Working Party wanted to remedy that by making the hub prominent as "both a *visual area* that invites people in *to experience its cultural offering* and a *collaborative hub* between renowned institutions . . . [to] draw in more visitors to this area and increase the exposure of, and enhance the quality of provision by, these renowned cultural institutions" (Cultural Hub Working Party Report to Policy and Resources Committee, 2013: 1, our italics).

As discussed in Chapter 2, culture-led regeneration is part and parcel of a redevelopment logic which leads cities simultaneously to transform their urban space and to develop unique place-differentiating brands, in order "to stay competitive in an international climate," to quote a Culture Mile manager. The same manager elaborated what this meant for the City of London and for the Culture Mile, namely being competitive meant having the right answers to questions like:

> what are the things valued by multinational organisations that are looking for new headquarters in Europe? What do they need, want and desire? And one of those key things was this sense of like a vibrant cultural, creative district. Like it's an exciting place to do work and do business. But also, it's kind of a good pull or draw for attracting world class talent.

We can see here clear parallels to the argument made by Florida that attracting creative workers and fostering the creative sector helps cities to become attractive and successful locations for investment (Florida, 2005; for a critique see Mould, 2015). Added to this is another Culture Mile aim, to diversify audiences and visitors while increasing the current footfall through the city:

> that idea of diversifying who engages with the district, but also thinking about our geographical position along the Elizabeth Line, Crossrail, when it finally open [. . .] trying to leverage the power of having a major cultural cluster in a part of the city and to transform the streets of walkways and lighting and so on to make it more of a destination, with an expectation to attract "an annual audience footprint of about 5 million". (Culture Mile manager)

The next step in the emergence of the Culture Mile was in 2014, when the City of London commissioned Publica, a private urban design practice, to develop a future vision for this cultural hub with the aim "to deliver a comprehensive identity for the area which will resonate and attract audiences from around London, the UK and the World" (Publica, 2015: 9). Given the arguments of this book, it should be no surprise that Publica's key recommendations consisted of a combination of *place branding*: including establishing a coordinated profile for the Culture Mile, a Culture Mile logo and a coordinated communication campaign about the area's cultural offerings; *urban design* suggestions such as developing a welcoming public realm and ensuring that the area is a well-connected, coherent, and navigable place; and *event management* proposals such as supporting a vibrant weekend culture by promoting cultural activities, street life, and café culture.

Following these recommendations, in 2016 the Cultural Hub Working Party requested £100,000 from the City of London to employ marketing and communication experts Jane Wentworth Associates and Pentagram to come up with a more distinctive brand strategy, which eventually set out four clear values for the area: joined-up, generous, agile, experimental. The Culture Mile name and imagery—logo, website, promo videos—also emerged from this work. The Culture Mile was officially launched in July 2017. Since the launch, it has been reported as a "genuine regeneration" (Pickford, 2017) which will create a "major destination" that will "deliver new experiences for everyone" (Kenyon, 2018). The Museum of London's relocation to Smithfield Market is being described as a landmark project that "will establish it with an international public" (Kenyon, 2018). The Culture Mile has also been integrated into the City of London's more recent 2018–22 Cultural Strategy. The new strategy continues to make a connection between the redesign of the area and cultural activities: the strategy's first objective is to "[t]ransform the City's public realm and physical infrastructure, making it a more open, distinct, welcoming and culturally vibrant destination" (City of London Cultural Strategy 2018–22, 2018: 9). This is being implemented by the Corporation of London's planning department through their "Look and Feel" program, which has started to apply the recommendations made by Publica for their public realm enhancement program (Figure 5.2). All this is seen, according to the chair of the City of London Policy Committee, as

> an essential step towards realising the physical transformation of Culture Mile. The area is currently dominated by vehicles and perceived by many to be hard to reach and difficult to navigate. These proposals set out an ambitious and exciting vision for the area which will create an unrivaled visitor experience, a welcoming environment for everyone to enjoy and a place where culture and learning are both created and consumed. (Zuccala, 2018)

All of these activities assume that the current state of the Culture Mile area, including Smithfield Market, is one that needs improving. Its current state is presented as hard to reach, difficult to navigate, not welcoming, not distinct, and not providing great experiences. This unattractive present is enacted by the particular version of the future proffered by the Culture Mile. Implicit in this future scenario building are the promises for a better kind of urban space. The Culture Mile strategy is an example of how "plans can be constructed to avoid undesirable futures, to make desired forecasts come true, or to create new, more desirable futures" (Myers and Kitsuse, 2000: 223). Indeed, this is another example of the storytelling described in Chapter 3.

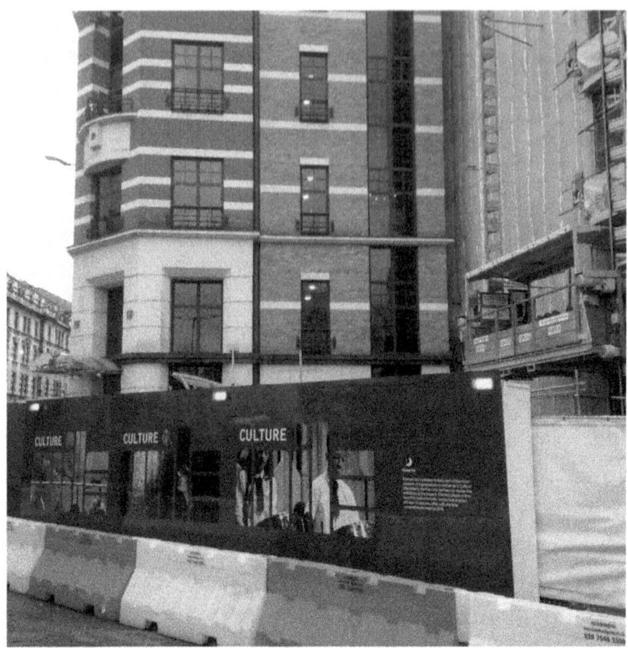

FIGURE 5.2 *A hoarding advertising the Culture Mile, Smithfield 2018. Image copyright Mónica Montserrat Degen.*

The Culture Mile regeneration project can therefore be understood as a form of anticipatory urbanism. In his discussion of futurities, Anderson (2010) insists that particular versions of the future always exist in a present moment: they are presented as an absence, something that has not yet happened, but there are myriad ways in which they can be represented, referred to, and even experienced. Indeed, Anderson argues that experiencing possible futures in the present is crucial to make this future happen: "[they] are constantly embodied, experienced, told, narrated and imagined, performed, wished, planned, (day)dreamed, symbolised and sensed" (2010: 783). We focus our attention in this chapter on the various practices which give content to a specific future for the Culture Mile and, in particular, for Smithfield Market's present, and how Instagram mediates these practices. In his work, Anderson (2010) identifies three key practices for mediating anticipated futures. First, practices of calculation refer to the ways in which the indeterminacy of the future is made more specific and calculable with numbers to provide certainty. This can be seen above in the expectations of how much economic income the urban transformation will generate in the future and how many visitors are expected to visit the area. Second, Anderson points to practices of imagination or "acts of creative fabulation," which can be visualizations or stories where "[m]aking the future present becomes a question of creating affectively imbued representations that move and mobilize" (Anderson, 2010: 785). Lastly, practices of performing the future that draw on embodied performances and make "futures present experientially. . . . The space of the exercise becomes an occasion for experiencing how a future event might feel" (Anderson, 2010: 786).

Much of the digital branding and urban activities of the Culture Mile project act to prepare the ground for future developments. By focusing on their temporal relations, we argue that we can understand the activities of the Culture Mile as an example of anticipatory urbanism in which "a future becomes cause and justification for some form of action in the here and now" (2010: 778). Throughout the rest of this chapter we particularly examine the last two practices of anticipation: fabulation and experiencing. We suggest that these practices produce a textured, lived urban aesthetic. And we emphasize how the digital mediation of these practices shapes a particular sense of the everyday, lived experiencing of Smithfield Market. Once place branding is mediated by a platform like Instagram, it becomes pervasively distributed and experienced. The next section explores the implications of that and suggests that such distribution enacts the *lived* aspects of a new urban aesthetic.

3 Instagram as an Expressive Infrastructure for Branding

Place branding is one of the key themes of this book. Brands are interpretive and aesthetic constructs. They work on an affective level "influencing people's ideas by forging particular emotional and psychological associations with a place" (Eshuis and Edwards, 2013: 1068). Place branding has become part of a widespread urban governance strategy to manage experiences and perceptions of places, "following the rationale that a place first decides what kind of brand it wants to become and then enhances developments to support that brand" (Eshuis and Edwards, 2013: 1067; see also Kavaratzis, 2009). As discussed in Chapters 2–4, and illustrated in the establishment and development of the Culture Mile in the previous section, place branding relies on creating a favorable image of place by producing or emphasizing certain functional, symbolic, and experiential aspects to attract visitors and investment.

Much of this urban branding is now undertaken by formalized private-public coalitions such as the Culture Mile which bring together the various interests from local government, businesses, real estate, and local non-for profit organizations. Indeed, there has been a clear co-evolution of urban redevelopment and branding, where the two processes work increasingly hand in hand and are part of urban policy "emerging as a hybrid materialization representing the process of creating new spatial settings" (Lucarelli, 2018: 12). During this process, the image of a neighborhood is an important element to be managed and shaped as it informs perceptions and expectations of a place. As the Culture Mile marketing manager explained, it took several attempts to get this right as part of the aim of the Culture Mile to reach audiences that usually do not access the key cultural institutions in the area: "how are we encouraging more people to change their perceptions and build new ideas around the Culture Mile and Culture Mile space leading to repeat visits and more kind of leisure time spent in the area?"

Place branding strategies are well known to utilize especially visual media to anticipate the feel and look of places. With the rise of social media in the last decade, place branding strategies have increasingly drawn on a range of digital platforms to disseminate their information and promote positive interpretations of the places they brand (Kavaratzis, 2009; Sevin, 2013). Most research has concentrated on the textual analysis of Twitter and Facebook posts (Andéhn et al., 2014). Instagram, however, is one of the most popular and least researched platforms. Instagram was first conceived in 2010 as a photo-sharing application. Nowadays Instagram is the fifth most used social network and the total number of photos shared has reached 40 billion with approximately 1.08 billion users and over 95 million daily posts (Smith, 2019). As a platform Instagram emphasizes the image, differentiating it from other networks and making it particularly relevant for us, as it reveals something about the users embodied point of view in place and their perception of places (Zappavigna, 2016). The affordances of the smartphone and Instagram app—taking photos or videos, applying filters, choosing which image to upload, adding a caption and hashtag—entails a series of aesthetic strategies and choices. Selectivity is a key feature of the history of photography, of course, but what is new is "that images can now be instantly uploaded and shared" and that the structure of Instagram, like other social media, enhances connections with other users using hashtags (Boy and Uitermark, 2017: 613). Hashtags demarcate topics and make themes or place experiences accessible to the public in "real time." Social media on a phone allows for "a form of visual co-presence arising out of the temporal nature of social streaming technologies, inflected by the portability of the mobile phone . . . a style of 'you could be here with me photography' . . . [inviting] the viewer themselves into the frame" (see also Leaver et al., 2020; Serafinelli, 2018; Zappavigna, 2016: 272).

The interactive nature of social media has transformed branding communication into a two-way process in which companies and customers enter into dialog. On Instagram corporations and organizations are able to engage with target audiences by following them, using hashtags and liking their posts: "Place branding and destination marketing projects, in this aspect, are no different from their corporate counterparts. It is no longer possible to separate place managers and public authorities as the sole producers of brand images and public as the consumers" (Sevin, 2013: 229). This has moved urban inhabitants from being just the target of place branding and marketing efforts, to also becoming producers. Because, of course, no places are blank slates for branding. In the case of Smithfield Market, there is already a strong sense of place in the area as well as a lively cultural scene, and the Culture Mile manager is well aware of this: "there is some understanding of what Smithfield is . . . [there is] awareness of it . . . the majority of Londoners . . . would have heard of it."

Social media such as Instagram thus enables residents or visitors to engage with place branding communication efforts, as the technology allows nonprofessionals to create and share images and text "which are assumed to provide a more authentic and innovative image of a place than professional photographs produced for strategic reasons" (Thelander and Cassinger, 2017: 6). User-generated content has become one of the most trusted and diverse source of online information for tourists; indeed, Instagram has become a "search engine" (Acuti et al., 2018; Boy and Uitermark, 2017: 617; see also Thelander and Cassinger, 2017). Acuti et al. (2018) state that "individuals are now actively looking for inspiration by exploring other consumers experiences, rather than expecting brands of more institutional sources to inspire them through traditional advertising" (Acuti et al., 2018: 190). Indeed, as their research on images of London and Paris on Instagram shows, "social media contributes to and accelerates place brand identity and brand image formation, facilitating and enabling stakeholders who constitute and live the brand in expressing their mental picture of a specific place" (Acuti et al., 2018: 201).

What our discussion so far highlights is that once place branding moves onto social media, it becomes embedded in a much wider set of communicative processes. It moves into many people's everyday experiencing of urban life. Chapter 4 made this point in relation to smartphone apps. While Chapter 4 focused mostly on the data generated by apps automatically, to generate perceived aestheticizations of urban space, here we are looking at how apps are part of the lived experience of urban space. In this context, users of space collaborate or are enrolled into branding processes but also post to capture their own experiences of place: "Instagram in as sense serves as a personalized brochure with appealing events and places" (Boy and Uitermark, 2017: 617). Indeed when we asked the Culture Mile marketing manager whether in his view social media is transforming placemaking, he told us:

> Yes, if you think about what people use to curate their own experience of an area, not just in terms of constructing outwardly that experience, but for themselves. It's not just about showing to other people that you've been to a place, it's also for you: how do you remember that place? What are the things you were really excited by? It absolutely has changed placemaking.

This presents a series of challenges for marketers as brands are increasingly co-created through everyday online conversations: "while a brand may initially embody a manufactured commercialized story, consumers' *storytelling* of personal experiences and opinions becomes absorbed into the brand narrative, hence changing, diluting or disintegrating its identity" (Lund et al., 2018: 271, our emphasis). Note the reference to storytelling again here. Storytelling is now "spreadable" (Jenkins et al., 2013), widely distributed among all sorts of social media users.

As we have emphasized throughout this book, sensory embodiments are mediated by digital technologies. CGIs and smartphone apps do this in different ways, as Chapters 3 and 4 discussed. In this chapter, we can see how embodied users of space are caught up in the lived experiencing of branding, making their own images, recording their own experiences, looking at the posts and views of others and caught up in those images of places. Carah and Shaul (2016) argue that Instagram (combined with the smartphone which it runs) can be viewed as an "attention capture and calibration device" (Wissinger, 2007: 235). However, Instagram mediates bodies in other ways too. Similar to the ways in which fashion photographers, models, and viewers affect each other by channeling attention through a careful choreographing of images, from posing in particular ways or editing the images, Instagram users similarly engage in aesthetic labor as "they make judgement about how to capture, edit and circulate images of their lived experience . . . [and] they watch flows of images and modify them by scrolling, liking and commenting" (Carah and Shaul, 2016: 71).

We now turn to examine this Instagrammed new urban aesthetic in more detail. We examine the images posted on Instagram with two hashtags: #SmithfieldMarket and #CultureMile. These hashtags are used by both the Culture Mile branding campaign and by the area's visitors, residents, and workers as they "explore the city and public spaces and, above all . . . communicate their experiences, also generating new images and experiences of it" (Toscano, 2017: 284). As we do this, we start to identify a third version of the new urban aesthetic. We explore how spreadable branding is generating a *dramatic* new urban aesthetic, and explain what that feels like in this particular area of London. We start by first examining the social media branding campaign by the Culture Mile.

4 Dramatizing the Culture Mile: The Culture Mile Branding Strategy on Instagram

The Culture Mile's social media branding operates via a range of platforms including a website, a YouTube channel, Facebook, Twitter, and Instagram. For the purpose of this chapter we examine the Culture Mile's Instagram account for its place branding, since Instagram is such an important platform for urban branding, as the introduction to this chapter noted. We start by examining the aesthetic labor that creates the Culture Mile's strategic branding aims online, analyzing what considerations underpin the Culture Mile's Instagram activity. We explore how place branding, social media posting, and in situ events are increasingly interlocked, such that Instagram activity is central to fabulating a *dramatic* new urban aesthetic which is staged in order to offer audiences exciting experiences.

The first Culture Mile Instagram post was on November 29, 2017, three months after the Culture Mile launch. It featured a nineteen-second panoramic video of the Culture Mile location and the logos of the Culture Mile logo and its core partners. By July 2020 the account had 158 posts and 2,399 followers: "postings are not huge yet," as the marketing manager admitted. Initially "there [was] not a longer term [social media] strategy for how that particular tool is used . . . but we've got clearer in the last year about how we use different aspects of social media." However, increasingly its social media campaigns "are [becoming] very important, we don't do much print at all, it's pretty limited. We obviously have a kind of hub website . . . we recognise that social media is becoming the first point of call for people to understand—to make choices, they're going to make more short term, there's going to be more impulse-based decisions" (Culture Mile manager).

Social media, then, are understood as immediate interventions into ongoing urban experience. While some posts feature historical information about the area or graphics advertising events, most of the Culture Mile posts feature images of events, installations, or exhibitions hosted by the Culture Mile or its cultural organizations supported by a written description. In-the-moment experiences are central in the Culture Mile engagement of audiences. The Culture Mile divides such experiences into three types: cultural, creative, and learning (Culture Mile London, 2020). These experiences are understood as an "animation of the whole neighborhood with imaginative collaborations and events . . . activated by exhibitions, gigs, pop-ups and events" (Culture Mile Brand Guidelines, 2020: 5). As discussed in Chapter 2, cultural experiences are crucial in the remaking of place identities during regeneration processes. Culture serves here as a catalyst to conceive of new interpretive grids by producing and mediating new sensations of place that influence the lived experience of place:

> the idea of using sensory experiences to slip something into somebody's day to day experience of the city, and of the Culture Mile area, is something that we're actively interested in pursuing, so we've done, you know, sound installations in the green spaces for instance. (Culture Mile manager)

While these activities involve careful planning over a long time, they are usually short, temporary pop-up events limited to a weekend, a night, or a few months and advertised in advance via social media (Figure 5.3). Increasingly these activities are conceived and curated in the physical urban space as "moments" that can be captured by a phone to be posted online:

> Around the Corner [the art] installation is basically taking people in a 'bread crumb trail' from Tate Modern, across London Bridge and up to Barbican. One of the key benefits of that particular proposal was that we knew it was going to get traction and pick up on social media . . . that sense of like using that sculpture, or one of the sculptures as part of the story of their experience of that place. (Culture Mile marketing manager)

This description of the Culture Mile on Instagram is a reminder of Böhme's use of the stage set as an analogy to understand aesthetic experiences. Chapter 2

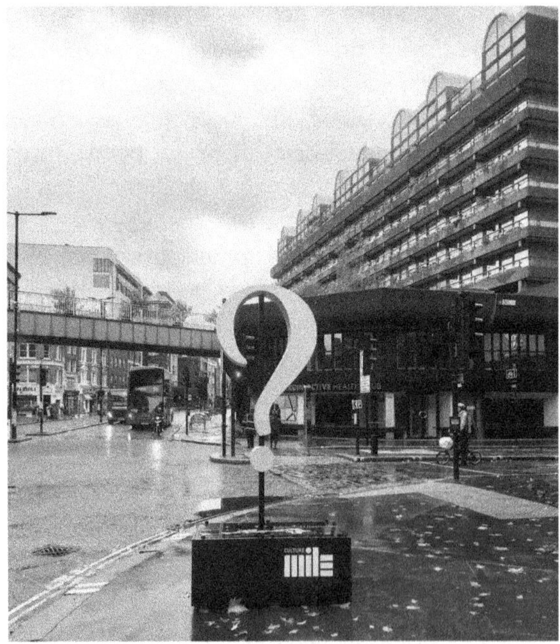

FIGURE 5.3 *An art installation in the Culture Mile, 2021. Image copyright Mónica Montserrat Degen.*

touched on his argument that "atmospheres are involved wherever something is being staged, wherever design is a factor [. . .] It is after all, the purpose of the stage set to provide the atmospheric background to the action, to attune the spectators to the theatrical performance and to provide the actors with a sounding board for what they present" (Böhme, 2013: 3). For Böhme, the stage requires certain "conditions in which the atmosphere appears" (Böhme, 2013: 3). Böhme draws on Plato to explain this form of atmospheric experience as a *phantastike techne*, meaning that an artist does not create art through mimesis but by consciously deviating from the original, taking into account the viewpoint of the observer; hence, "the artist does not see its actual goal in the production of an object or work of art, but in the imaginative idea the observer receives through the object" (Böhme, 2013: 5). Böhme offers the examples of impressionism or pointillism, where paintings do not aim "to copy an object or a landscape, but rather to awaken a particular impression, an experience on the onlooker" (Böhme, 2013: 5). In the Culture Mile, a sculpture or an event becomes part of an experience, one designed element among others that radiate out to combine and provide "something being staged"; but taking a photograph to capture that experience also positions the Instagram user as the observer attuned to what is presented. Coyne (2010) highlights how in an age of pervasive digital media and devices "tuning" is an apt metaphor to understand how technological devices such as the smartphone transform, recalibrate, and refine our relationships and experiences of places. Let's look at this in more detail.

When the marketing manager was asked whether he drew on a particular aesthetic for the Instagram account, he emphasized the "vibe," the embodied sensations that a post can evoke through its colors, design, and graphics. They do this by promoting experiences online: not just scenes, but also feelings, sensations, and emotions. Indeed, Instagram was conceived to capture the vanishing moments of everyday life, capturing the ephemeral nature of (urban) everyday life. Hence, the Culture Mile's branding on its Instagram account prepares its viewers for how a future event might feel. If "futures are created through the 'anticipatory experience' generated through both the acts of performance or play and the material organization of particular stages or sites" (Anderson, 2010: 786), then these posts stage the performance before it has happened. As the Culture Mile manager suggests, this happens

> [b]y giving people a sense of what they could expect. Because they might not necessarily have heard of Smithfield Street Party before, they might not necessarily have heard of Culture Mile before. There's a lot of things that we can't just rest on, that people will have a kind of base level understanding of. So like finding these shots, finding those kind of hero images, that like really compelling and exciting pieces of video content, that are going to communicate a sense of energy around place and excitement around

> place, [that] is really key. And that is something that we would do for every campaign across channels.

Since it is not actually possible to visualize events that have not happened, and since the Culture Mile has not yet organized very many events itself, the Culture Mile finds itself having to source appropriate images from other events. Rather like the staging of Msheireb Downtown in a warehouse in north London described in Chapter 3, Culture Mile has to stage the vibe of the Culture Mile future by sourcing images of past events which, as the marketing manager tells us, "were built on energy and reactions and excitement." These are then posted to brand forthcoming events and to anticipate the future feel of place:

> we kind of like found three of those and then made that like the cornerstone of our campaign across all channels. [. . .] Like this girl in the pink beanie [. . .] we were a bit like, that is the kind of person that we want to come to this event. That is the kind of thing we're expecting will happen during this event.

Note here how excitement is also anticipated for the bodies of Instagram users:

> [Physical space and online space] are absolutely connected. I think, for kind of nascent brand or a new place it is really important that you're clear about what you're trying to get people to engage with. So, across social media, this sense of curating may be a bit like reflecting a vibe that exists or reflecting the events that happen. Clearly showing a range of people that you want to attract to those events. All of that is really key through social media, because you're more compellingly showing, not just through words, what you can expect if you come here. (CM marketing manager)

This is close to the process of representing certain kinds of bodies that Chapter 3 described in the CGIs made for the Downtown Doha project. In their liveness though—their focus on events, on sensations, on the immediacy of the social media platform—this also speaks to the emergent sensibility of flow described in Chapter 4 (Figure 5.4).

This account of the Culture Mile's new urban aesthetic encourages us to describe that aesthetic as *dramatic*. It is dramatic because it is staged for an audience, as Böhme suggests. There is a relation between different things in the posted images, and between the images and the bodies that make and see them. But it is also dramatic because it is focused so often on intense and short-lived events: on heightened moments which puncture everyday experience. This might be an event staged by the Culture Mile, or it might be the moment of seeing something on Instagram which intrigues or excites. Dean (2010) argues that

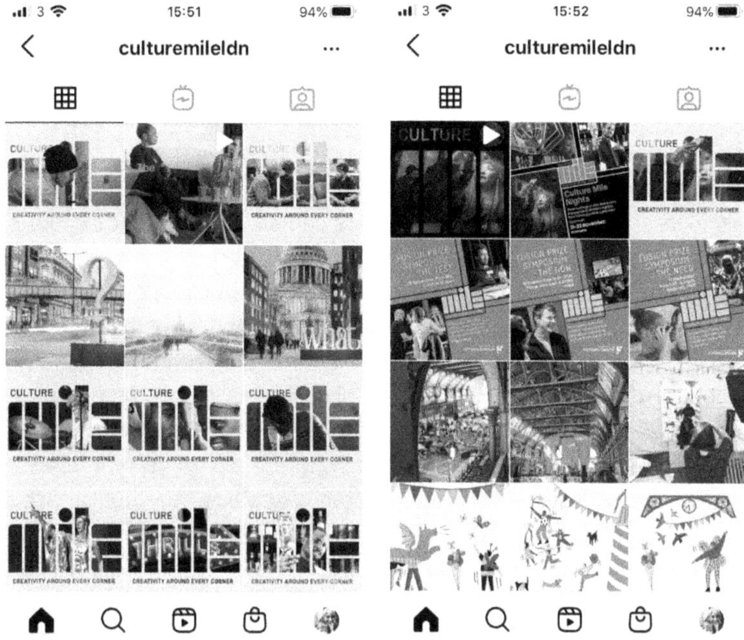

FIGURE 5.4 *Screenshots of the Culture Mile Instagram account, 2020.*

social media are encountered in a search for affective intensity—for something that will capture attention and provoke feeling. That so much of social media fails to do this only produces a search through more, so that "the active desire for jolts of interest, amusement, and diversion . . . drives users' restless motions across multiple screens and applications" (Paasonen, 2016). Finally, it is also dramatic because it is full of excitement and energy, thrills and reactions.

This is the third expression of the new urban aesthetic that we address in this book: the new urban aesthetic as dramatic. This new urban aesthetic assumes an audience, it performs excitement for that audience, and it intervenes in everyday routines with intense moments of energy both by offering events and also by puncturing ongoing social media feeds with offers of future stimulation. The next section explores how this dramatic new urban aesthetic plays out in Instagram posts made by people not laboring for the Culture Mile project.

5 The Lived Experiencing of Smithfield with Instagram

Chapters 3 and 4 both discussed digital technologies that require high levels of professional skill to design. The visualizers that create CGIs and the designers that create smartphone apps both have highly specialized skills and generally require high-performance hardware and software to carry out their aesthetic labor. Of course, the design of a platform like Instagram also requires specialist skills. But unlike the viewers of CGIs or the users of most smart city smartphone apps, the whole point of social media platforms is to share content among users (John, 2016). Since its release, Instagram has steadily increased what and how its users can share: from photographs with captions, to photographs with extended commentary, to videos, to stories, carousel posts (posts with more than one image), collections, more filters, direct messaging, and more (Leaver et al., 2020)—and what Instagram shares of the data it harvests from that activity has also shifted, most notably when it was bought by Facebook in 2012. Social media posts have thus joined the many ways in which cities are mediated; they are part of the urban imaginary of many city inhabitants. Following the accounts of influencers or organizations can deliver information about city places to your smartphone, as can searching for any number of hashtags. As discussed in Chapters 1 and 2, everyday practices in or encounters with places, whether as a worker or visitor, are textured performances in which the combination of the conceived, perceived, and lived moments of place amalgamate to produce particular and diverse forms of place engagements and experiences (Lefebvre, 1991). "These are mostly organic, bottom up processes, whereby places are claimed and shaped through everyday, and often mundane social practices" (Lew, 2017: 449). Thus, when people engage with places, they draw on social media posts, for example, to plan and inform their engagements, highlighting how sensory embodiments, urban imaginaries, and materiality are connected within personal moments and relationships to place. In this section we explore how the merging of in situ physical experiences and online images also evoke a dramatic new urban aesthetic, and how this reconfigures lived experience.

Before we analyze Instagram posts, though, we want briefly to sketch the character and atmosphere of the Smithfield area. This description is based on the survey in 2018 that we mentioned at the beginning of Chapter 1 which we undertook to examine the sense of place in the area (Degen and Lewis, 2020). The main characteristic of the physical space which the survey respondents emphasized was the juxtaposition of diverse sensory and temporal experiences, which creates a unique place identity in the Smithfield area. This stems in part from the built environment which features buildings from a diversity of historical periods. But the temporal juxtaposition within the built environment is intensified by a diversity of lived temporalities by various social groups whose distinct uses of the neighborhood generate particular daily and weekly rhythms and overlapping, clashing sensescapes in the public spaces of Smithfield. The use of public space varies greatly across the times of the day as different activities shape and conflict in space. From around

11:00 p.m., meat lorries start to arrive, loudly taking over roads creating traffic jams, and bringing in market workers in their white overalls shouting orders who jostle with young clubbers arriving and leaving the popular nightclub in the area. From about 2:00 a.m., a diverse group of market customers from exclusive restaurants to small halal butchers arrive at the market to buy fresh meat. And from around 8:00 a.m. till early evening, Smithfield's streets are full of a mixture of finance and creative workers mingling with builders, hospital staff, couriers, tourists, as well as visitors to local restaurants at nights and weekends (see Figure 5.5).

These overlapping, contrasting, and intense social uses of the public spaces gives the area a very distinctive feel, which one of our interviewees describes as follows:

> It's got a bit of bite to it . . . it's a bit grit, a bit of edge, a bit trendy here and there . . . This place is alive at 3am in the morning. . . . You've got weird little cocktail bars around the corner underground. There's the market just around the corner. It's all this mash of . . . it just feels it's got that little edge to it, which is interesting.

All of our survey respondents commented on how it is precisely the variety of social uses that give the area a particular idiosyncratic and vibrant character. One local resident we interviewed described Smithfield as "buzzy" and commented on how unexpected encounters were commonplace, as different groups found themselves "cheek by jowl" in this area of the city. The Wordcloud in Figure 5.6 summarizes their descriptions of the Smithfield area.

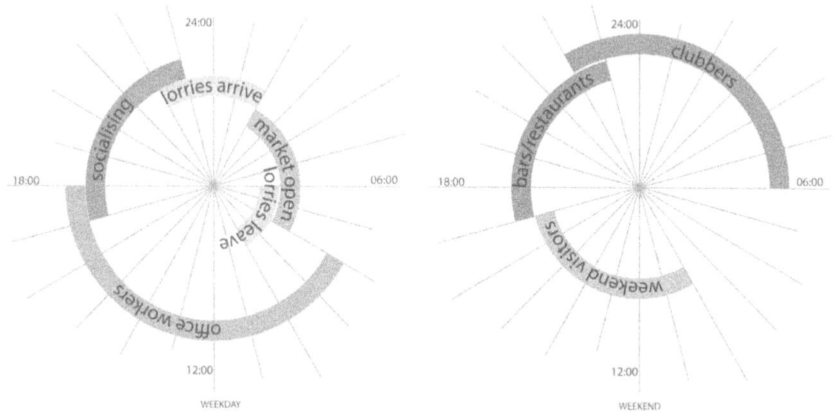

FIGURE 5.5 *The fluctuating and overlapping temporal rhythms of the Smithfield Market area, 2017. Image copyright Mónica Montserrat Degen, designed by Isobel Ward.*

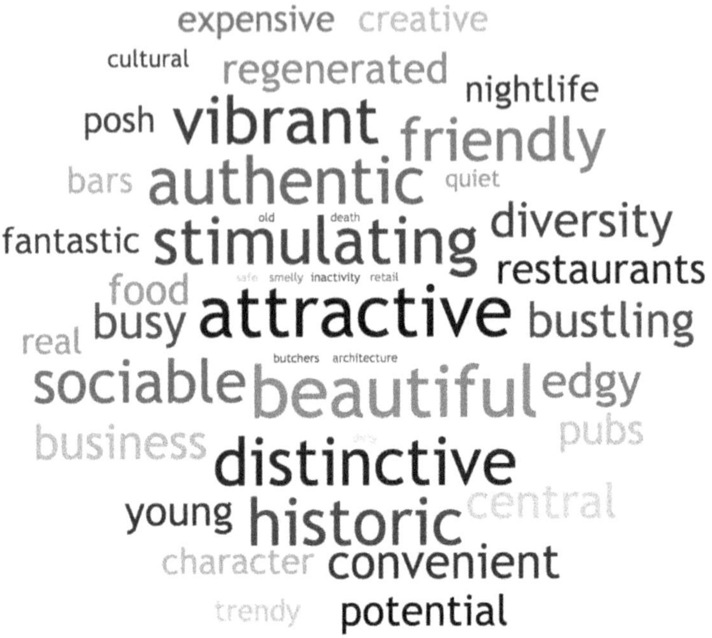

FIGURE 5.6 *Wordcloud describing the feel of place in the Smithfield Market and surrounding area, 2017. Image copyright Mónica Montserrat Degen, designed by Isobel Ward.*

By no means all of the people doing these different things in Smithfield will be posting about them on social media. As the Culture Mile marketing manager told us, Instagram is good for "curating that [event] and making it look really appealing for a certain audience . . . we knew that Instagram [would attract] an audience between 18-35, late thirties, they were like the key audience for it" (Boy and Uitermark, 2017; Leaver et al., 2020). Nonetheless, many will be. And so too will some of the visitors to the various Culture Mile events and activities that the chapter has discussed in Sections 1 and 3. We now turn to what people post on Instagram using the hashtags #CultureMile and #SmithfieldMarket. Our qualitative analysis is based on a sample of 500 posts of all posts which up to September 2019 had ever used those two hashtags (a total of around 10,000). The discussion in Sections 3 and 4 of this chapter suggests that there should be some overlap between the sorts of things shared by the Culture Mile team with what other users share. What our analysis shows is that #CultureMile and #SmithfieldMarket posts display quite different content but they do share an emphasis on the dramatic.

Let us start by discussing what was photographed and posted with each hashtag. Table 5.1 summarizes our findings. Photographs tagged with #SmithfieldMarket focused heavily on the architecture of the Market, and on

Table 5.1 The Percentage of Instagram Images Featuring Different Kinds of Content

#CultureMile		#SmithfieldMarket	
Joy and Peace artwork	20.2%	views of the Market building	10.7%
Tunnel Visions light installation	18.3%	food and drink in bar or restaurant settings	9.7%
art/music/sculpture/exhibition	17.6%	cast iron architecture	8.8%
individual person	9.6%	fresh meat	7.5%
garden or flowers	8.2%	the Market's Grand Avenue	6.6%
the Barbican Estate	7.5%	a plaque or building sign	6.6%
advertising information	7.5%	four traditional red postboxes	6.5%

meat being sold in the Market, and on food and drink in bar or restaurant settings. The photographs tagged with #CultureMile were much closer to the branding campaign run by the Culture Mile team although only 1.5 percent of the posts analyzed were directly posted by the Culture Mile. They focus heavily on art and music events and their advertising, with some attention given to the green spaces and the Barbican housing estate in the surrounding urban environment. Interestingly, given the Culture Mile's emphasis on people experiencing the excitement of the Culture Mile, there are a greater proportion of photographs of individual people in images with the Culture Mile hashtag than with the Smithfield Market hashtag. Figures 5.7 and 5.8 show what we have selected as the most typical of each of these types of photographs. The size of each image in the figure reflects the image dominance in our sample. For example, the colorful installation in the center of Figure 5.7 is an artwork that featured in over 20 percent of the #CultureMile images, and so is the dominant image in the figure.

Figures 5.9 and 5.10 visualize the most common words used in the image captions in our sample of posts. There is also a divergence between the words used in relation to the two tags, broadly reflecting the difference in the image content. The text in #CultureMile posts, like the images, refer to art and music events or organizations: art, installation, streetart, performance events, temporary, and words connected to art installations recur, as do color, music, sound, and lighting. The Barbican Centre, the London Symphony Orchestra, and the Guildhall School of Music—partner organizations of the Culture Mile—are also mentioned. However, the captions also mention Smithfield Market and the Barbican. #SmithfieldMarket photographs' captions, meanwhile, focus heavily on place names and food. As well as Smithfield there are other London locations: a restaurant, the Barbican,

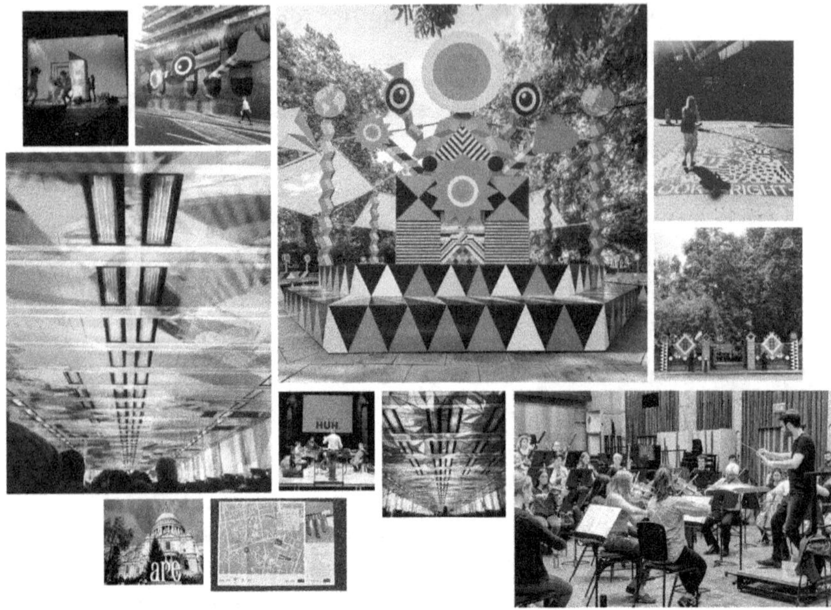

FIGURE 5.7 *Imagecloud of #CultureMile Instagram posts, in order of frequency of posting: "Joy and Peace" art installation, "Tunnel Visions" art installation, "art/music" events, individual people, elements of nature/gardens, the Barbican Estate, informational materials for Culture Mile, miscellaneous people, Beech Street Tunnel, "Around the Corner" art installation, and a group of people. Image copyright Mónica Montserrat Degen, designed by Jennifer Dodsworth.*

Billingsgate, Farringdon, and the Charterhouse complex of buildings (all near the Market), Battersea and Clapham, as well as various permutations on the Market's name, and London itself. Food-related words in captions include meat and beef (as well as butchers), cocktails, gin, wine, beer pub, glass, chefs—and yummy and foodporn. There are also one or two captions that refer to events: Christmas, lunch. Art, streetphotography, and blackandwhite also appear.

Both hashtags refer to stagings of the same urban environment, but they focus on very different features. It is clear that Culture Mile has been successful in branding its events and activities as the posts using its hashtags all contain events, street art or cultural events related to the Culture Mile. Indeed, all the photographs of Smithfield Market that appear with the Culture Mile hashtag are of events held in the Market building (see Figure 5.11). As the Culture Mile marketing team intended, this creates a temporary sense of place built from intensely experienced moments rather than ordinary everyday life routines. The sense of drama and excitement that the Culture Mile project wants to generate parallels the edgy buzz that many users of

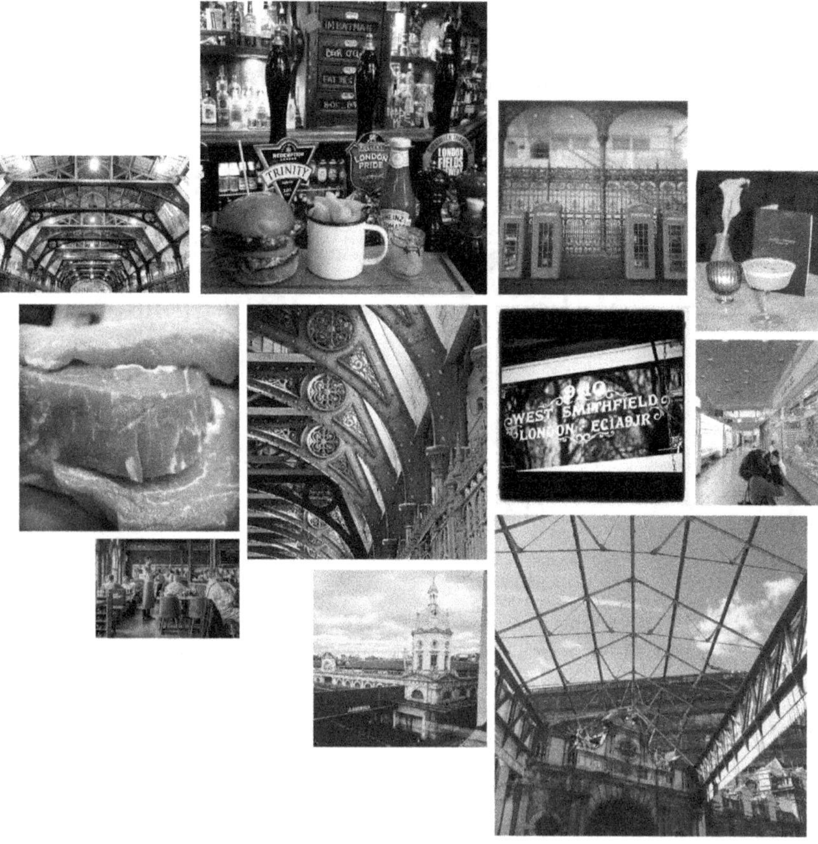

FIGURE 5.8 *Imagecloud of #SmithfieldMarket Instagram posts, in order of frequency of posting: Smithfield Market, Smithfield market roof, individual person, cold drinks, and an interior of a café, restaurant, or bar. Image copyright Mónica Montserrat Degen, designed by Jennifer Dodsworth.*

Smithfield Market describe. Both generate drama from temporary moments of juxtaposition.

The #SmithfieldMarket images, in contrast, are heavily focused on the materialities of the urban environment: food, meat, architecture, and buildings. People appear much more rarely. If Instagram's portability and digitality have contributed to the "hyper-representation of the world (e.g. from the photo of the cappuccino in the morning to the Friday night out with friends) showing the predominance of visual elements in many daily practices" (Serafinelli, 2018: 2), these are rarely just casual snaps. Many of these images are highly crafted, deliberately emphasizing sensory qualities like taste, pattern, texture, and color. Looking more carefully at the Instagrammed images of Smithfield Market also suggests a kind of intensity, but this time a visual and spatial one.

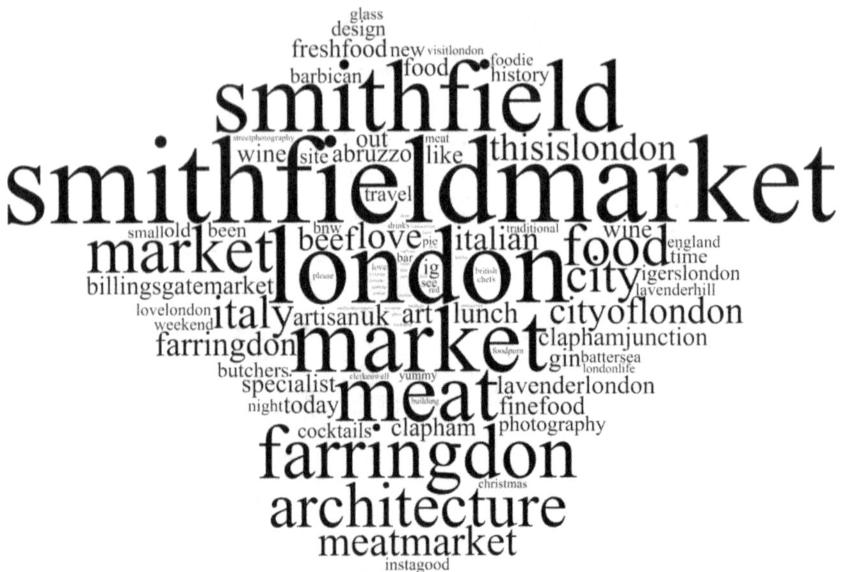

FIGURE 5.9 *Wordcloud of #SmithfieldMarket Instagram post captions. Image copyright Mónica Montserrat Degen, designed by Jennifer Dodsworth.*

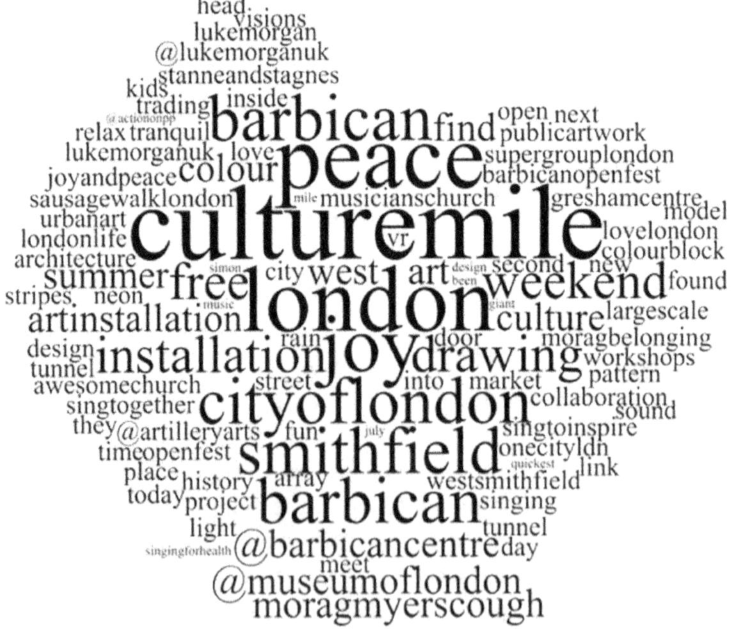

FIGURE 5.10 *Wordcloud of #CultureMile Instagram post captions. Image copyright Mónica Montserrat Degen, designed by Jennifer Dodsworth.*

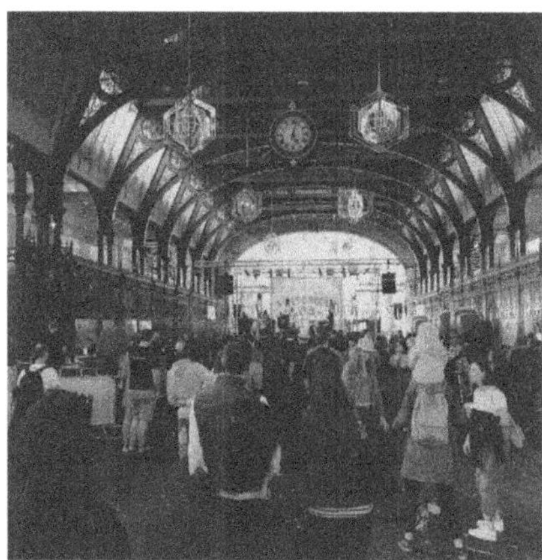

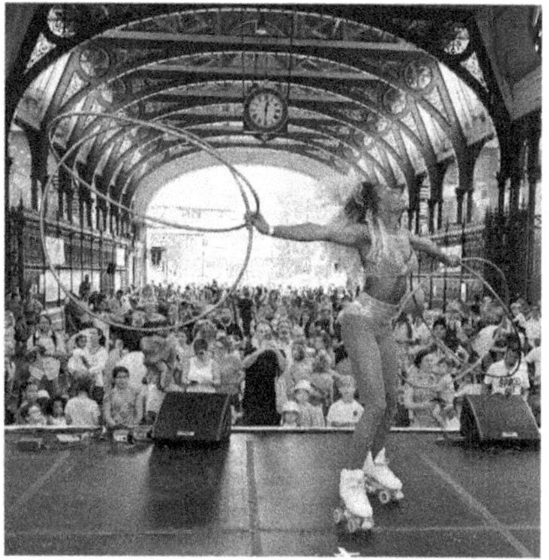

FIGURE 5.11 *Two Instagram images of an event at Smithfield Market tagged with #culturemile.*

Photographs are carefully framed, cropped, and enhanced using Instagram's editing tools, to create a specific visual aesthetic.

This aesthetic has changed over time. When Instagram launched, its retro filters were wildly popular. They invoked older photographic film effects, and offered the "look of aged photographs as they appear to us in the present" (Bartholeyns 2014, 60 quoted in Zappavigna, 2016: 286). Many commentators remarked that this retro look made the edited photographs heightened or dramatic moments of expression "rendering more poignant the present moment" (Zappavigna, 2016: 286) and adding the "emotional value provided by a temporal distance that is made visible by a dated aesthetic" (Luders et al., 2010, 960 quoted in Zappavigna, 2016: 287). Figure 5.12 shows two photographs of a favorite #SmithfieldMarket Insta topic, a set of six old public phone boxes in the Market. (This topic itself plays to nostalgia—these phone boxes are now rare—as does the popularity of old street signs in the Market posts.) The phone box photograph taken in 2011 clearly shows the faded, slightly sour, saturated retro color palette so popular when Instagram launched. The other was posted in 2018. The retro look is no longer quite so popular, and the colors of the postboxes are now brighter and richer. What remains though is a careful attention to composition: to the arrangement of things in the square of the Instagram photograph. #SmithfieldMarket photographs often focus on tightly cropped images of architectural details, meals, and drink and fresh meat (see Figure 5.13). The images that picture market scenes are also usually strongly organized by the geometry of the architecture. This is a different kind of drama from the temporal juxtapositions emphasized by the people in Smithfield and by the Culture Mile marketing team. Still shared with an audience, this is a spatial drama of scale and detail, bringing sensory

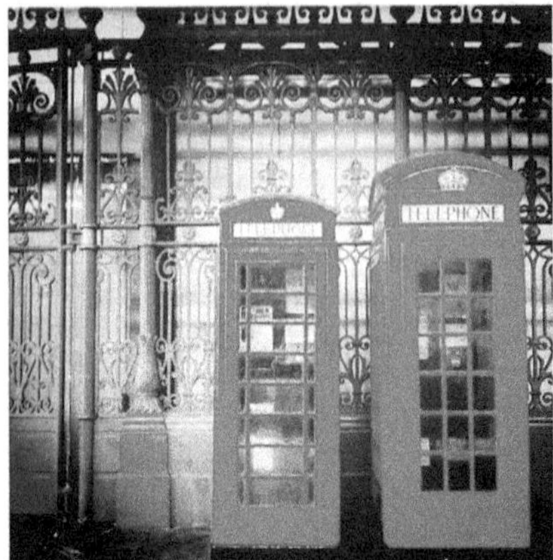

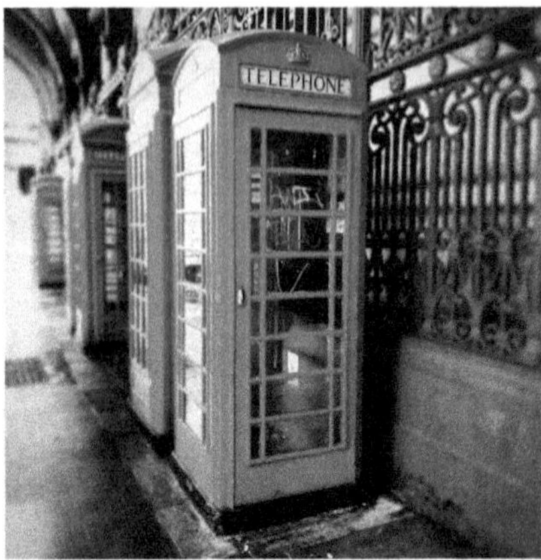

FIGURE 5.12 *Two Instagram images of the phone boxes at Smithfield Market tagged with #SmithfieldMarket. The one on the left posted in 2011 has a blurry retro feel and dimmed colouring, the one on the right was taken in 2018 and shows a sharper focus and a different use of filters.*

details up close, giving a surprising point of view, or an unexpected crop (see Figures 5.13 and 5.14).

In a recent study on the relationship between digital outdoor advertising, bodies, and urban space, Dekeyser states that digital technology "enables an intensified capacity to selectively open up to, integrate and interact with urban spatiality" (Dekeyser, 2018: 2). We argue that this is also the case with digital social platforms, especially Instagram. As others have noted, new embodied sensibilities and new ways of looking and engaging are emerging with Instagram. From this discussion, we can see that one way in which Instagram is recalibrating lived urban experiences is by attuning our realms of urban perception toward dramatically aestheticized moments: snapping a performance, pausing on something gorgeous on your Insta feed, or crafting the perfect Insta post: "Instagram users train their eye to spot slices of the world around them worthy embalming" (Boy and Uitermark, 2017: 616; Coyne, 2010). Such moments are staged, time-limited, and exciting, and they feel like that whether we are urban branding or on a day out. Urban spaces are designed and intensified to feel like exceptional interventions into everyday experiences of time and space.

The dramatic new urban aesthetic thus consists of interventions into the current moment, both generating and capturing some kind of sensory intensity, which is then shared with a social media audience. As we noted, for the branding workers, this is about creating a vibe that will contribute to an anticipated urban future, one that will see the end of Smithfield Market currently experienced. But for both the branding agency and the visitors, mediating the exciting moment in this way is also about rendering the urban

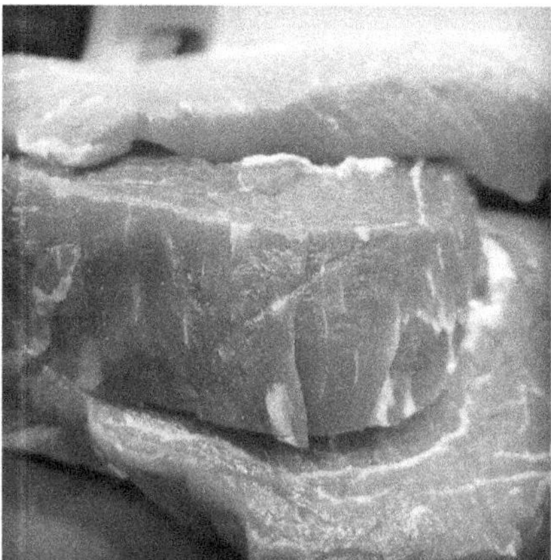
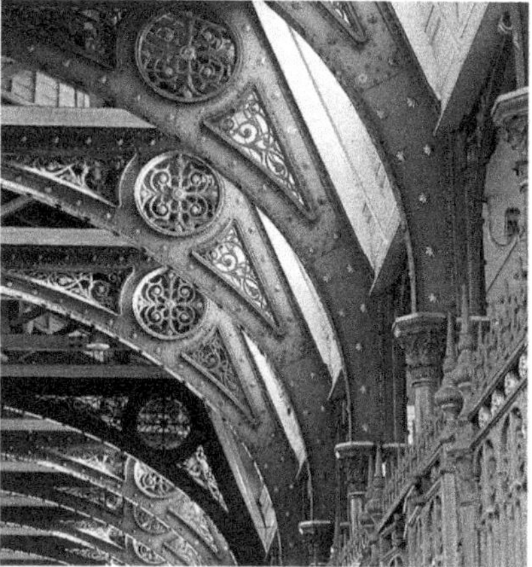

FIGURE 5.13 *Impactful details from #SmithfieldMarket.*

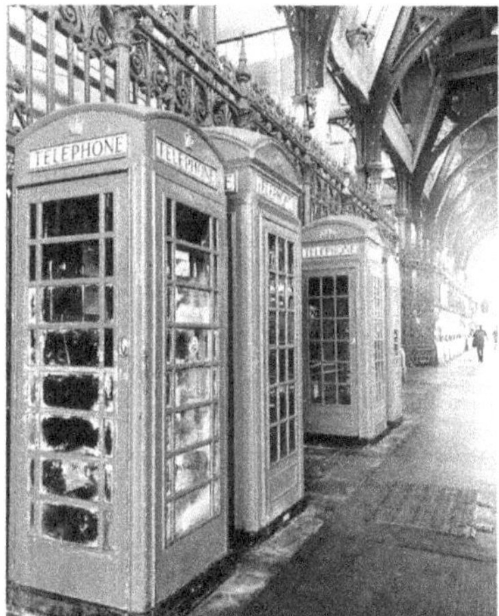

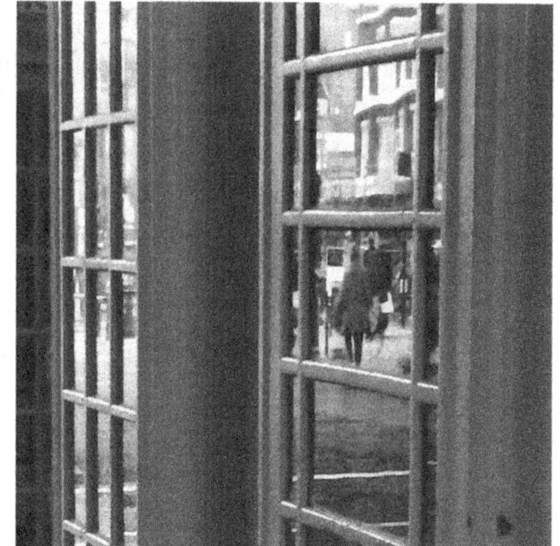

FIGURE 5.14 *Careful staging of the #SmithfieldMarket phone boxes.*

environment into a memorable space in the future. As Culture Mile's marketing manager says:

> It certainly is something that we consider when we look at public programming: how is this going to be remembered? How is it going to be distributed or shared? It's not the number one guiding principle but it's absolutely something that is at the back of my mind. For example, if you are doing a Smithfield Street party, there's so much stuff that is happening . . . but from a marketing and a press perspective, it's always really key to know what those kind of big visual wow moments are. Because for legacy

> and like making sure that you capture the essence of the event, that's extremely important. But also making sure that people have that kind of one image on one piece of video content, that they are like: "Wow, do you remember when. . . ."

So Culture Mile events are not only about generating "big visual wow moments" in the present that are shared; they are also about ensuring that even when the event is in the past it will continue to generate a wow: "wow, do you remember when." The drama of the new urban aesthetic then, with the help of Instagram, interrupts the present but also promises future excitement too.

6 Conclusion: Texturizing Drama

To conclude this chapter, let us briefly summarize and reflect what our case study tells us about this third, dramatic version of the new urban aesthetic and its textures. Our case study of a culture-led regeneration project, the Culture Mile, was another clear example of the urban branding we have discussed throughout this book. We highlighted how this regeneration project brings together coordinated strategies of place branding, urban design, and event management not only to represent and reorganize the spatial organization of places but also to reorganize the embodied experiences and sensations associated with this branded place, an increasingly common strategy in contemporary cities (Degen, 2010; Degen and Rose, 2012). In discussing this case study, we were particularly interested in how it was mediated through a digital social media platform: in this case, Instagram. It is important to point out here that the increasing use of social media like Instagram as part of urban branding campaigns can also be understood within the framework of platform capitalism. Leaver, Highfield and Abidin (2020) expertly unpick the ways that the Instagram platform and its users are implicated in value extraction in multiple ways (see also Boy and Uitermark, 2017; Hearn and Banet-Weiser, 2020). It is also the case that the users of the Smithfield currently perceive the area in terms of manifold, overlapping, and juxtaposed social and temporal uses of place which have not been entirely colonized by digital mediations. However, there are also other kinds of power relations at work here. Urban branding is more and more often designed to curate the physical environment and cultural activities such that they can be shared on social media platforms like Instagram in highly aestheticized ways. The increasingly pervasive and various uses of social media platforms are mediating "the communicative weave, or fabric, that is created through human activities in space" (Jansson, 2013: 286) and thus the sensory feel of particular spaces. It is these sensory shifts that we have attended to here.

From our discussion in this chapter, we can see once again that the new urban aesthetic enacts a textured kind of urban spatiality. The chapter has touched on some of the conceived elements of the Culture Mile project such as the careful planning of a placemaking strategy by an urban design practice. The use of Instagram is also about perceived space, too, in the way that the platform's smartphone app is used to stage images of urban sites that capture and share an experience. But the chapter has especially emphasized how these two aspects are bound together with the lived aspects of digitally mediated urban space. The lived aspect of triadic urban space is what saturates everyday life and holds it together as a normal way of doing things (Jansson, 2013). As images are routinely made and shared,

> a form of visual co-presence is arising out of the temporal nature of social streaming technologies, inflected by the portability of mobile media. . . . "you could be here with me" photography . . . visual sharing can approach synchronous time. In other words, images, delivered either to massive online audiences or to smaller networks, can be posted or viewed in what is referred to as "real-time." (Zappavigna, 2016: 272)

We have argued that from a temporal perspective we can understand these real-time processes also as the enactment in the present of an "anticipatory urbanism." We follow Anderson's (2010) suggestion to analyze the relation between the "presence of the future" in the present moment and its dynamics in a "living present," and examined the role digital mediations play in how urban environments are made through a constant folding of futures into the present. We identified the use of practices of conceived fabulation, such as the visualizing of future events on Instagram and other digital platforms, and lived practices of performing the future which work on the corporeal sensations the future might bring, such as the programming of events and cultural activities in the Smithfield area, as ways in which the future feel of the Culture Mile is conveyed. Our aim has been "to understand how the experience of the future relates to the materiality of the medium through which it is made present, whether that be a graph or an affective atmosphere" (Anderson, 2010: 793). The Culture Mile project offers a specific kind of future for this area of London, and the aesthetics of Instagram in the present moment align with the excitement and intensity promised for the future.

In the age of social media, branding and placemaking operate as a distributed practice where destination marketing organizations no longer have a monopoly on the ability to structure or define meanings of the fabulation of future places but have to work with other users who are also posting on platforms like Instagram: "Brands and brand identities are co-created in an active negotiation of organisational and consumer identities" (Lund et al., 2018: 276).

For example, in the case of the Culture Mile, we were told of the importance that one "influencer" had in posting and attracting visitors to one of the Smithfield Streetparty events. The Culture Mile marketing manager identified an

> influential family blogger . . . lots of families with younger children are following other parents, who are giving them suggestions. And you could see that [the event], once it happened once, there were a couple of people that had a following [in their] thousands. It was amazing! And the second weekend there were so many more people who came . . . and that big shift in attendance was to do with the amount of traction we had on social media through people coming and be like: "This is great, you need to come back. It is happening next week."

This distributed sharing is central to what the chapter has characterized as a *dramatic* version of the new urban aesthetic. It is dramatic because the shared images are carefully staged, with different elements filtered and edited into stylized Instagram posts. Instagram enables a staged, costumed, and intensified version of everyday lived experience (Böhme, 2003). It is dramatic too because these images are made for an audience. The Instagram user is entrained in the scene by their desire to scroll through others' posts. It is also dramatic because what is posted is a sensorially intense and time-limited experience. It is an exceptional event or sensory encounter, like a temporary art installation or a glistening hunk of meat, which interrupts the everyday. We have identified differences in the visual content of what is posted on Instagram by the Culture Mile marketing team and people photographing Smithfield Market. Nonetheless, the lived uses and effects of social media platforms like Instagram converge with the Culture Mile marketing strategy to become a kind of new normal for this area of London. While the #CultureMile Instagram posts tended to picture art and music, #SmithfieldMarket posts were of food and striking architecture. In both cases, though, the lived aspect of triadic urban space was dramatized. The dramatic new urban aesthetic, mediated as it is by one of the biggest social media platforms, becomes a pervasive sensation of lived urban experience.

6

The New Urban Aesthetic and Its Power

1 Introduction

Through case studies of urban change in Milton Keynes, Doha, and London, we have illustrated some of the ways in which the mediations of digital technologies are radically transforming how urban spaces are conceived, perceived, and lived (Lefebvre, 1991). This transformation is happening partly because of ongoing efforts to redevelop and rebrand urban spaces. It is also part of another intensifying process: namely the increased digital mediation of the urban. All three case studies demonstrate that various digital technologies play an important part in the re-organization of urban space into sensory atmospheres. Computer-generated images, app-generated data circulations, and social media imagery all intervene in how urban space is textured and therefore how it feels. Given the importance of sensory feelings to this digital mediation of urban space, we have described this transformation as *the new urban aesthetic.*

The book has identified three versions of the new urban aesthetic which it has described as aesthetics of *glamour*, *flow*, and *drama*. Chapter 3 analyzed how CGIs were used to guide a large urban redevelopment in Doha, Qatar. The particular aesthetic evoked in these CGIs is one of luminous glamour: acts of digital manipulation allure audiences in sensory ways in order to make "'ordinary' objects, surfaces or people appear magical, elegant, effortless and establish their difference from the mundane" (Huopalainen, 2019: 333). In Chapter 4 we analyzed how the city of Milton Keynes is branding itself as a smart city and also how it is generating new urban experiences through smart city apps. Here, the new urban aesthetic feels like a sense of flow, of constant movement, a feeling mediated through the app-facilitated mobility of bodies and data. Then Chapter 5 looked at a third manifestation of the new urban aesthetic: the online place branding of the Culture Mile in London. By comparing those corporate branding practices to everyday lived experiences of place and the postings of other Instagram users, we showed how the social media platform stages a dramatized urban aesthetic of temporally circumscribed, sensorially intense, staged experiences. Glamour, flow, and drama are all ways in which sensory embodiments are entrained through digital mediations into feeling things in urban spaces.

These three chapters have demonstrated what were key themes in Chapter 2. First, they demonstrate that the new urban aesthetic is configured differently in different places. Each of our case studies discussed different forms of urban redevelopment and branding strategies: the Msheireb project is a large-scale redevelopment project, Milton Keynes is about retrofitting an existing city, and the Culture Mile is a cultural regeneration project. In each we explored different media: CGIs, smartphone apps, and a social media platform, and explored how each is producing specific textured mediations. Second, each chapter also focused on what we have termed aesthetic labor, which is the work done by corporeal bodies to mediate urban spaces aesthetically. Thus, we examined the work done by the professional designers, architects, and visualizers involved in the Doha redevelopment project as well as the efforts of a place branding organization on Smithfield and app designers in Milton Keynes. We also positioned Instagram users as everyday aesthetic laborers in the Smithfield area of London. The Milton Keynes case study looked at app users too, and emphasized how the data circulations of their digital embodiment worked to produce certain sensations. Thus, we highlighted how aesthetic labor results in a broad range of different experiences of digitally mediated space. Third, each chapter also emphasized a different aspect of the spatiality of the new urban aesthetic to show how they are differently textured. Chapter 3 stressed the *conceived* aspects of a new urban aesthetic: its abstractions, representations, and categorizations. Chapter 4 examined a new urban aesthetic as *perceived*: its shaping by the materiality of digital devices and software. Chapter 5 explored a new urban aesthetic as *lived* in normalized everyday practices. Conceived, perceived, and lived are terms that we have adopted, like so many other scholars, from the work of Lefebvre. While our use of them is faithful more to Lefebvre's phenomenological leanings than his Marxist commitments (Kincaid, 2020; Simonsen, 2005), we nonetheless feel they are helpful in pointing to forms of differentiation in the new urban aesthetic. As Lefebvre emphasized, actually every space has its conceived, lived, and perceived elements and in that sense, our emphasis on just one in relation to each of our case studies is artificial. Nonetheless, it allows us to specify different versions of the new urban aesthetic. This chapter focuses on our final move to embed differentiation into the conception of the new urban aesthetic, namely to attend to the "distribution of the sensible" (Rancière, 2006).

Our emphasis on difference is driven by our concern to address directly the way that new urban aesthetics are powerful. They produce particular kinds of urban spaces and urban futures (and indeed presents and pasts). We therefore need a nuanced critical vocabulary that can begin to describe their specificities and enable the staging of other kinds of aesthetics. Kincaid (2020), for example, offers a reading of Lefebvre that insists that the lived experience of urban spatialities and temporalities can be challenged if the triadic experience of space does not cohere. What happens when certain kinds of bodies are not represented or perceptible in CGIs of glamorous urban futures, for example? Drawing on intersectional feminist work, Simonsen describes this conflict:

> The additional "task" faced by men and women of colour is one of reconciling their own "tactile, vestibular, kinaesthetic, and visual" experiences with the operation of a "historical-racial schema" producing "racial parameters" within which their corporeal schema is supposed to fit. (Simonsen, 2010: 231)

Minoritized bodies disproportionately suffer this alienation, this "crisis in their involvement in the world" (Simonsen, 2013: 20). The triadic quality of space may offer resources for critique of such crises because, as Kinkaid highlights, Lefebvre's theorization of social space "operates as a challenge to the homogenising spatial logic of capital and bourgeois culture, threatening to unravel its ideology and produce space *otherwise*" (2020, 168, emphasis in original). So, for example, the various forms of aesthetic labor we have outlined may work to achieve different kinds of mediations. They might challenge established representations or perceptions of place by altering, ignoring, or resisting them.

This chapter deepens further our understanding of the new urban aesthetic as differentiated. It does so by returning to the work of Rancière and his notion of the distribution of the sensible. We develop a critical vocabulary which can attend to the production of distinctive spatial-temporal forms of urban sensibilities while *also* marking the way that each production displaces other possible configurations (see also Sumartojo and Pink, 2019). Chapter 1 remarked that every version of the new urban aesthetic is always accompanied by "the loss of mediations never to be actualized" (Kember and Zylinska, 2012: 19). This chapter focuses on describing the three versions of the new urban aesthetic in ways that allow those losses to be marked. Previous chapters have explored what is produced as the new urban aesthetic. Here, instead, we pay attention to what is rendered marginal or impossible. This chapter is therefore an attempt to redistribute the sensible in relation to these case studies: to see and sense their aesthetics differently, to become attuned to other possible aesthetics, to redistribute their sensibilities so as to render these urban transformations differently.

The first term in our critical vocabulary is *storytelling*. We introduced this term in Chapter 3 where we discuss the conceived urban aesthetic of the Msheireb Downtown development by focusing on the forty-two computer-generated images that were selected to visualize the master plan. Storytelling was one of the ways in which visualizers and architects described how they conceived the work that their CGIs were doing. The second term is *animation*. Animation was introduced in Chapter 4 as a way of referring to the emergent qualities of digital mediation, which Chapter 4 discussed both in the context of data circulations in smart cities and in the circulations of digital visualization about smart cities. The third term is *seamfulness*. Seamfulness allows us to pay attention to aesthetic labor. The CGIs of urban redevelopment schemes and smart cities discussed in Chapters 3 and 4 are the result of production processes that can take place in many different locations. That labor is not obviously visible in the final images, which appear seamless and picture-perfect. Seamfulness, as a critical term, attempts to reconfigure the distribution of visibility to make that invisible labor available to perception.

Each of the case studies focused mostly on just one of these critical terms. However, we propose that they can be used to review any version of the new urban aesthetic. Seamfulness, for example, can pick at the divisions that structure all kinds of aesthetic labor. As we will see in Section 4 of this chapter, it can address the work of editing an Instagram post as much as the production chains that create CGIs. And we assume that other studies of other versions of the new urban aesthetic will require and invent other terms to describe the consequences of those specific distributions of the sensible.

All three terms—storytelling, animation, and seamfulness—are ways of describing how different configurations of the new urban aesthetics are organized and put into practice. They offer some ways of describing and feeling the times and spaces of urban aesthetics, and they also allow the critical description of their effects. They offer ways to explore how the production of scenes, nostalgia, invisibility, viewpoints, futurity, and so on also produce exclusions, erasures, futurelessness. They suggest different kinds of critique of the new urban aesthetic. The next section briefly reprises our conceptual framework for understanding the new urban aesthetic before elaborating on our critical vocabulary.

2 Power and the New Urban Aesthetic: Differentiating and Distributing

A key aim of this book has been to emphasize that the new urban aesthetic manifests in very different ways. It is different in different places: it emerges through different processes and it enrolls different kinds of corporeal experiencing. More than this, the book has also emphasized that these various manifestations of the new urban aesthetic are powerful. They have effects. They make some city spaces feel glamorous, seductive, and luxurious; others feel exciting, with events and installations dramatically interrupting the everyday; others feel as if all you have to do is move with your screen to experience new information, new efficiencies, new flow. As these aesthetics emerge, different digital embodiments are also configured: laboring bodies, bodies represented in particular ways, bodies assembled from data flows.

Chapter 2 outlined our conceptual framework for understanding this differentiated new urban aesthetic. It began with outlining the work of Böhme on atmospheres to make sense of the emphasis of contemporary capitalism and urban development on experiences. Böhme (2014; 2017) highlights the conscious production of atmospheres, particularly relevant in processes of urban redevelopment, and emphasizes the aesthetic labor involved in creating atmosphere. Objects designed to evoke atmospheres need to be positioned, organized, sensuously choreographed, and displayed in particular arrangements to reach out and address a spectator, as we have argued throughout our case studies. Throughout the book, our emphasis on the aesthetic as a sensory experiencing has encouraged us to think about how corporeal bodies are caught up in different kinds of urban aesthetics. In this sense, our emphasis is different from Shanti Sumartojo and Sarah Pink (2019), who suggest that individuals encounter what they call urban atmospheres, and that a "diversity of experience at the individual level remains" (Sumartojo and Pink, 2019: 122). We have suggested something a little different and show how through at least some digital mediations, embodiment itself is reconfigured.

Nonetheless, we share Sumartojo and Pink's concern to consider the power dynamics of different aesthetics and atmospheres, and the argument in Chapter 2 began to work toward that by thinking about the different components of a Böhme-esque atmosphere. We turned to Lefebvre's account of the social production of space to understand how atmospheres are conceived, perceived, and lived when mediatized by digital technologies. That move has helped us "to conceptualize the sociospatial [and temporal] significance of mediatization at different levels" (Jansson, 2013: 282). Following Jansson (2013), we emphasized Lefebvre's understanding of a "textured" production of triadic urban space and time which sets the "practico-sensory body" at the center of his theorizations. Hence, theorizing how space is produced involves a focus on bodily experience as it emphasizes "the body's implication in and constitution of a 'sensory-sensual space'" (Simonsen, 2005: 1). The sensory body here is both a mediator and active agent in experiencing and shaping the coexisting, concording, and interfering relationship between the three spatial elements.

Hence the book's turn to Rancière for an understanding of the power of the new urban aesthetic as a *distribution of the sensible*. As Chapter 2 outlined,

Rancière's interest in the distribution of the sensible seems particularly appropriate for describing the consequences of the convergence of urban branding with digital technologies. His notion of the distribution of textured times and spaces allows us to specify their sensory effects. It also allows us to understand how particular distributions render certain sensibilities unrecognizable. Rancière uses the term "police" to describe the institutions which enact a particular distribution of the sensible: these can literally be the police but can also include all the other forms of state power as well as other organizations and practices that establish and normalize particular distributions of what is sayable, visible, audible, and sensible. The "police" achieve "the configuration of a perceptive field" (Dikec, 2015: 92); they are "a heterogeneous set of organizational forms, technologies, and strategies for ordering, distributing, and allocating people, things, and functions" (Karaliotas and Swyngedouw, 2019: 374).

> They draft maps of the visible, trajectories between the visible and the sayable, relationships between modes of being, modes of doing, and modes of doing and making. They define variations of sensible intensities, perceptions, and the abilities of bodies. (Rancière, 2006: 35)

However, Rancière does not straightforwardly line up the police with dominant organizations or institutions. Indeed, our case studies have identified a range of actors who police distributions of the sensible more or less deliberately. City branding agencies, for example, or CGI visualizers, work at creating particular aesthetic experiences by making certain kinds of images, or designing certain kinds of events or social media campaigns. App developers also assume particular kinds of urban inhabitants: mobile and dynamic bodies. But we have also suggested that everyday uses of digital technologies also map visibilities, spatialities, and temporalities. There are conventions for picturing places on a platform like Instagram, for example: particular things that are more photograph-able and share-able, particular compositional features like colors, cropping, and scale, and which are achieved by careful aesthetic labor.

This approach to thinking about power relations and the new urban aesthetic also generates a particular understanding of "resistance." For Rancière, resistance is always possible because people occupy multiple and complex positions in various distributions of the sensible, and bring those experiences and perceptions together in inventive ways (see also Rose, 2017). In his book *The Emancipated Spectator*, he argues that

> there are only ever individuals, plotting their own paths in the forest of things, acts and signs that confront and surround them. The collective power of spectators does not stem from the fact that they are members of

> a collective body or from some specific form of interactivity. It is the power each of them has to translate what she perceives in her own way, to link it to the unique intellectual adventure that makes her similar to all the rest in as much as this adventure is not like any other. This shared power of the equality of intelligence links individuals [. . .] This capacity is exercised through irreducible distances; it is exercised by the unpredictable interplay of associations and dissociations. (Rancière, 2009: 16–17)

It is the case that Rancière relies here on a rather unproblematized notion of the "individual"—but his sense that there is always an "unpredictable interplay of associations and dissociations" is an important reminder that the policing of distributions of the sensible is always contestable (and see Dikec, 2015). We now turn to the critical terms discussed in Chapter 3—storytelling, animation, and seamfulness—to elaborate how that might happen.

3 Storytelling and the New Urban Aesthetic

As discussed in Chapter 2, the development of a new urban aesthetic is driven by two interacting processes. First, the intensification of branded urban redevelopment schemes across the globe. This intensification refers both to the increased number of developments across the globe (Datta and Shaban, 2016; Swyngedouw et al., 2002; Weber, 2010) and to the increasing speed with which they occur (Chien and Woodworth, 2018; Datta, 2016; Raco et al., 2008). The financing of such future developments requires powerful branding strategies that create enticing new urban imaginaries that will sell these future places not yet built. To stand out, these urban imaginaries need to suggest, evoke, and then deliver on the ground, unique and aesthetically engaging destinations. Secondly, digital technologies provide the means to both create and distribute data both in cities and about cities. The data then becomes part of many different devices which mediate urban spaces, from smart city control center dashboards to CGIs to smartphone apps and many more.

Chapter 3 focused at length on a specific set of digital images generated to sell and brand an urban redevelopment project in Doha. A key term that the digital visualizers and the Architectural Language Advisor used repeatedly in describing their work was "storytelling." For example, the ALA's definition of the "Memorable Moment" he wanted the CGIs to evoke was "a slice of life which carries a story and also a resonance" such as "a bike, blurred, in motion" or "a child, running or smiling, towards the camera." Similarly, a visualizer told us:

> that's what all these images are about, nothing else really. The key to the image is that you are telling a story, whether the story is: "This is what this building will look like on this particular day with this particular lighting condition and that's how the interior behaves and how the light will fall." Or whether your story is "This is the garden area that Qatari locals will go with their families."

Indeed, storytelling is at the center of many product marketing strategies. Storytelling is increasingly considered as a strategic branding tool for urban places as they engage audiences on an emotional and experiential level, indeed it is a "central tool for communicating the experiential value of place" (Bassano et al., 2019: 10). Stories are a key component of urban imaginaries more broadly. All our chapters have shown how particular stories provide interpretive grids which try to convincingly frame the experience of future urban developments, whether glamorous, flowing, or dramatic. Specific versions of the present and future are offered. Digital technologies enhance storytelling devices in particular ways. Let's look at examples from our case studies.

As just noted, in Chapter 3's case study of Msheireb Downtown, storytelling was one of the ways in which visualizers and architects described what the CGIs they crafted were doing. They wanted viewers to be captivated through embodied experiences by immersing them in street-level images, walking or cycling, to co-experience the mood of the CGI and an aesthetic of global chic. Storytelling is part of these effects in a number of ways. They organize the temporality of the event being depicted as well as organizing future expectations by audiences. They are one of the most important ways in which the future is made present (Anderson, 2010).

One of the most important of these organizing narratives in Doha was the story of a massive transformation into a new future state. "Stories are devices . . . to making sense of ongoing change interpreting and reinterpreting the present looking at the past and the future" (Bassano et al., 2019: 10). Each CGI image invites the viewer to participate in a journey in an unfolding story of a particular urban future. CGI viewers also experienced other spatiotemporal experiences of the development and entrains the body of the viewer, for example by being able to "see" the same places at different times of the day, or feel the different rhythms of public life depicted: a family gently strolling down a mall or animated businessmen crossing a square for a meeting. The CGIs have to depict a believable storyline, by reflecting an accurate temporal daily order and experience of the redevelopment. Storytelling consists of a new form of place branding consistent with the experience economy as it "allows territories to tell stories which arouse special atmospheres and involve real or future clients emotionally, which increases their sense of territorial belonging and identification" (Bassano et al., 2019: 13). In all these ways, storytelling becomes a strategic device in an aesthetic economy "whenever 'place' is conceived as a service system and in building a competitive identity to obtain reputation through value cocreation processes" (Bassano et al., 2019: 11).

Storytelling, then, pitches a particular version of a place and its futures as it silences or conceals others. To turn this branding technique into a critical term, we want to highlight that there is never just one story about a city. Any place can be storied in very different ways, thus its conceived imaginary can be multiple. This is very clear in the case of Msheireb Downtown. The development was built on the site of a tight-knit neighborhood with low-rise buildings that housed low-income immigrants mostly from Kerala, India, for over fifty years: "an ethnicized landscape of disrepair, decay, and poor living standards emerged at the heart of the historic city, which the Msheireb Downtown scheme has been designed to replace" (Melhuish et al., 2016: 233). Neither those former inhabitants of the site nor indeed the migrant workers who built the new development are featured in the developer's marketing materials.

Storytelling is also central to the representation and conception of smart cities as discussed in Chapter 4. In smart cities all sorts of images are digitally produced and distributed, they can be part of city branding campaigns or part of city management techniques. Storytelling is important here in the way that the smart city is made credible through discursive work and visualizations that depict and evidence smart activities and smart urban environments. Again, these stories often stress the future of the city, telling stories about how life will be different once these new technologies are widespread. Many smart city promotional videos align the ability to tell stories about smart cities with particular embodiments. Many videos have talking-head interviews, and these are usually with either with officials or elected representatives of city councils who have adopted smart technologies, or with senior employees of the corporation making the video: almost all these technocrats are men, and mostly white (Rose, 2018).

Moreover, only some corporealities are aligned with particular anticipated smart urban sensoria. Stories about smart cities persistently align the use of urban smart tech with white masculine professional bodies. Videos demonstrating the use of CAVs are typical here. In the 2014 animation about CAVs in MK produced by DVC, the point of view is the user of the CAV, who lifts white, male hand holding smartphone, where a message arrives informing that the pod is ready. He presses a button on the screen of his smartphone and the fighter plane-like canopy of the pod lifts and invites him to sit. When inside, the calm-sounding female narrator informs us that while he is being chauffeured to his next destination, he may use his "valuable extra time" and the pod's inbuilt screen to catch up on emails, read newspapers, or play games. The style and gendering of the voice invites the same criticisms that have been made of digital assistants in using a female voice for a subordinate service role (Fessler, 2018; Strengers and Kennedy, 2020). A study of how driverless cars are advertised to the general public by Hildebrand and Sheller (2018) shows similar gendered patterns. The marginalization of women as active agents is a persistent feature of the CAV images. Moreover, all the bodies pictured in the MK CAV visuals are dressed in business suits. The emphasis here is clearly on CAVs as an aid to service sector productivity. Much of the promotional imagery made by corporations selling smart city technologies and by cities showing themselves as smart is conventional. If children are pictured, they are almost always accompanied by a woman; only women shop, or cook. If you're old, it generally means you're pictured as in need of smart healthcare, usually delivered by women. Many of the videos also picture only white bodies (though that is beginning to change in more recent videos, particularly from the United States). In terms of who is represented as leading smart cities and benefiting from their technologies, then, the picture is clear: professional, mostly white adults, mostly men. The alignment of only some bodies with specific kinds of smart city activity is a clear example of a distribution of the sensible.

Lastly, in Chapter 5, storytelling was also a prominent feature of the ways in which social media posts by diverse users of places become part of place branding and narratives of cultural redevelopment strategies such as the Culture Mile in Smithfield, London. Storytelling becomes a key device of the branding strategy of place, becoming a formula. For the Culture Mile manager, their own work was part of a narrative of London redevelopment projects, one that was speeding up: "the shorthand is to understand that what took fifteen years in Shoreditch took seven years in Haggerston, takes five years in Catford, takes three years in Peckham, you know, will potentially happen overnight because now everybody knows the formula." As this quotation highlights, storytelling has a fabulatory role in urban redevelopment in that, as Anderson (2010) argues, it is an anticipatory action through which the future is made present: "a set of possible (rather than probable) 'as if' geographies are made present through forms of visualization and narrativization" (Anderson, 2010: 785). However, social media also multiply these fabulations, as official strategies and the general public posting hashtags and images of places emerge, converge, and diverge. "Digital media, in fact, contributes to transform storytelling in a collective and dynamic process that reinforces members' escalating engagement with the community itself" (Bassano et al., 2019: 19). The immediacy of social media ensures that stories can be posted and modified quickly and seamlessly; stories can be nonlinear and social media users can co-create a story by blending it with their own everyday temporal practices and expectations (see also Bassano et al., 2019).

At this point, it is possible to note a close alignment between the conceived elements of the spaces of the new urban aesthetic and the work of storytelling. The conceived aspects of space are the abstract representations that tell or show something about that space, and as they do so, that space is given a certain kind of value. In the stories told in Doha, Milton Keynes, and Smithfield, the future that those places are moving into is positively valued. Thus, in the case of the Smithfield area regeneration, newspapers announce "the potential" (Rogers, 2018) of the Culture Mile—that will "deliver new experiences for everyone" (Kenyon, 2018). But of course these are highly partial stories about the future, in both senses of the word. They are both limited and judgmental. A crucial aspect of our use of the term storytelling as part of a critical vocabulary is that it allows us to identify that partiality. The focus of these stories of urban change and transformation is only on a particular vision of the future. It values just one story about one future (and therefore by implication about one present, too, the moment of change). But that timeline and frame of experience is just one from an array of possibilities that architects and developers and marketing agencies draw from. These are the "police," in this case, to use Rancière's term: urban professionals like architects, developers, and branding agencies, for whom this conceived future will bring positive outcomes.

Our Smithfield case study suggested that it is not possible to make a simple distinction between such professionals and those with other relations to urban redevelopment, however. Branded images of the Culture Mile intersect with those of Smithfield, both in terms of what is pictured on Instagram and how it is pictured. Both places are in part about drama, and that drama is intensified and circulated via Instagram by many different people. The practice of using Instagram, then, is also a kind of policing: it stabilizes that place into a particular look and feel (see also Boy and Uitermark, 2017). But this is a popular policing as it enables displaying the self in relation to everyday practices in the city: "Instagram users [to] act out aesthetic and lifestyle ideals as they craft images and strategically display aspects of their lifeworlds" (Boy and Uitermark, 2017: 622).

Hence, thinking about the new urban aesthetic as in part a kind of storytelling prompts critical questions about policing, selectivity, and representation such as: Who is telling what kinds of stories? Whose stories are heard? Which bodies are visible in the visualizations of future city life and which are not? What are they doing? Who is in the narrative and who is forgotten? Who and what is present in the story's future and who and what are left out? Sometimes the answers to these questions are clear-cut. The driverless cars in MK are not being shown with passengers who have disabilities or who are mums with three toddlers and ten shopping bags. In the Culture Mile case study the 24-hour greasy spoon cafés which support the work of the market workers are not featured; neither are the meat market workers or local residents. But in other situations, the answers might be less straightforward.

4 Animating the New Urban Aesthetic

We also want to find a somewhat different critical vocabulary which speaks to the aesthetics of extractive data harvesting and analysis. As Munster (2006) elaborates at length, the data circulating through digitally mediated cities remains entangled in corporeal sensibilities. For example, Chapter 4 argued that the data circulations anticipated by a handful of smart city apps—no matter what their content referred to, whether breastfeeding or cycling—all assumed a data-driven, efficiency-seeking mobile body. This body can be understood as the apps' user: the corporeal body carrying a phone and using its many apps. But this body is also datafied. It is co-constituted with data flows, emerging as part of data circulations. It is an incarnation of the doubled digital embodiment described by Munster (2006: 23) as "a kind of graft, which is an unequivocal mark of connection and difference" between the fleshy body and datafied body.

Munster's account of digital embodiment thus emphasizes both the corporeal body and its sensory experiences as well as its mediations by data. If "new media entice bodies to venture towards incorporeal flows of information and combine, in convergent and divergent ways, the capacities and functions of carbon materialities with those of information flows" (Munster, 2006: 23), she does not merge the corporeal and the datified—there is connection but also divergence between them. She is therefore particularly interested in the specific qualities of the digital and thus to the consequences of the fact that "our bodies are immanently open to these kinds of technically symbiotic transformations" (Munster, 2006: 25). It is Munster's emphasis on the forms that emerge from the virtual that interests us. It aligns with other accounts of networked data which also try to evoke in more sensory terms the feel of data and thus give attention to "forms and processes of becoming; to assemble and be assembled; to tune forms and formations" (Redstrom and Wiltse, 2018: 161). As data circulates and accumulates, what might be imagined is

> a stratified constellation of technical memory matter, composed of resources that shape political and cultural imaginaries. This stratification should not be thought of merely as across, but also in terms of depth, height, scale, extensiveness and duration. . . moving in different directions. . . . Its forms

> may change and its content migrate, accruing or shedding textures in the process. (Withers, 2015: 17–18)

For Withers (2015: 22), this multiplicity opens up possibilities for other orientations to and in the spatial and the temporal.

Chapter 4 introduced the term "animation" to refer to this shifting, labile quality of the new urban aesthetic of flow. Data is gathered at scale and as it circulates and is processed in all sorts of ways, more and more data is generated: "data's power to give rise to more data" always generates a surplus (Clough, 2018: 129). This constant replenishment and modification leaves the possibility for "open up other spatial and temporal orientations, and harbour potential to communicate alternative forms of . . . experience" (Withers, 2015: 22). The emergent qualities of data always contain the possibility of transformation, recalculation, mutation, and refiguration. Algorithmic analyses "articulate and disassemble populations in real time" (Clough, 2018: 113). It is this quality of transformation that the notion of *animation* emphasizes (Levitt, 2018; Manovich, 2016).

There are many ways in which the potential of (digital) animation to assemble things differently is stabilized. In the case of CGIs showing street scenes composed of many different digital-visual elements, Chapter 3 noted the explicit rules that architects and visualizers were asked to follow so that their models of specific buildings could be integrated, and that the development and its future inhabitants looked appropriately Qatari. Chapter 4 emphasized the consistent assumption across very many different kinds of smart city apps that mobility was normal and desirable, should be as efficient as possible, and only required appropriate information to be triggered. Chapter 5 also pointed to shared assumptions about cultural regeneration and the role of social media in articulating a seductively dramatic version of a new urban future.

However, all of our case studies offer examples of how the assembling and disassembling of digitally mediated embodiment produced surplus and excess, of mutability and emergence: of animation. In the case of the Instagram posts of Smithfield Market and the Culture Mile, this concept of animation adds another layer to the multiplicity of #Smithfieldmarket's that exist online. For example, as well as the multiple stories, multiple sensory events puncture the hashtag feed, receding as another emerges in a flow of proffered sensation. And as we realized during our analysis of Instagram posts multiple versions of #Smithfieldmarket (from a market in Belfast to a suburb in Sydney) appear online, sometimes creating confusion about what and where is being experienced.

Being alert to the processes of animation allows us to think about other aspects of the Doha redevelopment project too. In that example, the unlimited ability to change CGIs resulted in an extraordinary proliferation of image files as thousands were produced. A digital image can be tinkered with endlessly, generating a huge number of different versions. There are also immense numbers of different versions of the highly elaborated images. Both of these things posed challenges to the digital infrastructure which hosted the Doha CGIs. Version control became exceptionally important, and some files ended up being too large to open on some visualizers' desktop computers. Emergent proliferation can be overwhelming. With more files and more versions and more screens and servers comes the risk of technical failure. Moreover, once the CGIs reached a certain state of finish, they started to appear in a range of less-anticipated places: in the curriculum vitae of freelance visualizers, for example. There they became much less about Qatari cultural identity and much more about demonstrating highly skilled aesthetic labor. And in principle, once let loose from the Msheireb Downtown projects, those CGIs could have been manipulated and reproduced in any number of other contexts, transforming their meaning and aesthetic once again.

Chapter 4 argued that only mobile bodies seem to be valued in the smart city (Rose et al., 2021) and that is their distribution of the sensible: mobile bodies. But what also emerges alongside mobile bodies as they are constantly generated are bodies that are not mobile. Even as a certain kind of default mobility and rationality was being assumed and produced by the apps' design, so other sensoria were also brought into being: bodies that were immobile, irrational, inefficient. Bodies that did not enact this disposition toward mobility were seen as a problem that had to be fixed. The development of the prototype Age Friendly Map app for the Theatre District of Milton Keynes provides an illustration of the constitution this less mobile—less smart—embodiment. The Age Friendly Map began as project between MK Gallery, Age UK Milton Keynes, and local visualization studio DVC, when the Gallery realized a key demographic was missing in their outreach and engagement. As the Age UK worker explained:

> I know when we were working with Milton Keynes MK Gallery, they were talking about hitting people within target ages. So they wanted to reach people between the 55 and 74. So that is, technically, the younger older person. So they have, generally speaking, more stable income, their health is better, they recently retired or might be taking part in activities, and so they're people who want to get out there.

This description of the "target" demographic as people "who want to get out there," who want to be mobile, and so the Age Friendly Map app was designed with this younger-older-but-potentially-mobile body in mind.

> And so we sat around a table and we just, kind of, literally, off the tops of our heads, we thought, "Well, people might want to know this, and they might

> want to know that," and so in our heads it was just, I suppose, like a map that highlighted certain places where they could get information. But . . . you can't overload this, because if you look at a standard map, with road signs and directions and everything, there is a lot of information but it's always presented in the most simple way, because that's the most effective way.

For the intended users, it was assumed that simplicity was key. This was a particular concern for the app designer. It was assumed that this "demographic" required certain kinds of information to get them moving to the city center—special offers for seniors in shops and cafés, and the location of public toilets and benches—but that they would find using an app difficult. In effect, these "younger older bodies" were being constituted as less technologically able, and as poor, easily tired and less continent than other bodies. Shadowing the mobile embodiment, then, is the possibility of immobile embodiment and the work of inducing more mobility into existence.

In considering the various "human-data assemblages" (Lupton, 2018: 1) convened by the new urban aesthetic, then, it is important to remember their animatory potential to shift, recombine, reemerge, and mutate.

5 Seamfulness: Seeing Aesthetic Labor

Much of the storytelling about digital technologies emphasizes their speed and their smooth integration. As we saw in MK, discussion about smart cities emphasizes their efficiencies and their savings of time and labor; in platform urbanism, the flow of data in a smart city needs to be plentiful, uninterrupted, shared in real time, and integrated. Platforms' apps should feel smooth (Leszczynski, 2019); CGIs are meant to look seamless, and even magical, as if they are snapshots of the future. A dramatic new urban aesthetic erases the question of labor differently again, by appearing suddenly and surprisingly, apparently without preparation. While it is particularly important for glamour of any kind to appear effortless (Thrift, 2008), the aesthetics of flow and drama also minimize effort in their apparent speed and integration. But of course, all aesthetic effects are produced by many kinds of labor; indeed, Chapter 3 suggested that that labor always produces more labor because the emergent qualities of digital data need constant stabilization. People work with different kinds of hardware and software to produce different elements of all the distributions of the sensible that this book has described.

> Far from increasing the feeling of dematerialization, digital techniques have rematerialized the whole chain of production. Today it is impossible to ignore that, whenever a printed [image] is available, there exists, upstream

> as well as downstream, a long and costly chain of production that requires people, skills, energy, software and institutions and on which the constantly changing quality of data always depends. (November et al., 2010: 584)

Chapter 2 drew on the work of Böhme to emphasize the aesthetic labor that goes into the making of the many material mediations of the new urban aesthetic. Aesthetic labor "designates the totality of those activities which aim to give an appearance to things and people, cities and landscapes, to endow them with an aura, to lend them an atmosphere, or to generate an atmosphere in ensembles" (Böhme, 2003: 72). As our case studies have suggested, this labor is extensive and diverse.

Geographers and others have begun to give considerable attention to the many kinds of labor that enable platform urbanism, smart cities, and digital visualizations to happen. Some studies have focused on the highly skilled work of architects and visualizers who use various design softwares (Houdart, 2008; Rose et al., 2014; Yaneva, 2009); others have explored the workers who labor on the digital-visual effects seen in so many movies now (Chung, 2018), or those who design computer games (Ash, 2010) or smartphone apps (Cockayne, 2016). Considerable attention has also been paid to the somewhat less-skilled workers who find work using outsourcing platforms, or who do routine design or visualization tasks. In both cases, different kinds of work take place in different locations. Projects are broken down into many tasks, and each task is sent to where the necessary labor costs are least. "The increasing digitisation of work and recent advancements in automation and communication technologies don't just augment the labour process with digital data, digital processes, and machines; they also embed it in stretched-out networks of production: with tasks quickly passed in complex assemblages from person to person, person to machine, and machine to machine" (Graham and Anwar, 2019: 2).

A limitation of the critical vocabulary of storytelling and animation that we have developed so far is that they do not allow the labor that goes into the making of say CGIs to become apparent. We now introduce a third critical term that allows us to discuss labor as part of the new urban aesthetic. That term is *seamfulness*. Like storytelling and animation, "seamfulness" asks us to reorient our immersion in the new urban aesthetic and to consider the effects of its particular distribution of sensations and sensibilities. Anyone involved in the design of digital interfaces and databases is very familiar with the complexities of their production, and we suggest to look at new urban aesthetics in order to see the work that has gone into making them. We want to perceive "the material residues of globally dispersed workforces and digital production pipelines in the aesthetic forms of contemporary [cities]" (Chung, 2018: 8; and see Chalmers, 2003). As Chung (2018) discusses, a film's digital-visual effects entail lots of different digital elements which are merged together in any one image—just like CGIs. Chung works on identifying those multiple components in a movie and then on identifying where they were worked on and by who. Her visual and spatial sensibility thus focuses less on smooth integration and more on layers, joins, and seams: on "gaps, ridges, and wrinkles. . . trajectories. . . textures" (Chung, 2018: 33). (And we too have been using Jansson's (2013) term "texture" to describe triadic urban spaces.) The point of her analysis is to consider how "each digitally manipulated layer is materially connected to specific sites of production and how the composited layers retain the spectral, yet perceptible, residues of a geographically dispersed workforce" (Chung, 2018: 7). Seeing the components that are brought together to make a finished image is thus a way of acknowledging the labor of that compositing and integration. It reconfigures the spatial and temporal organization of a new urban aesthetic's distribution of the sensible.

Let us follow Chung's methodology and map the locations of (some of) the work that went into making the Msheireb Downtown CGIs. The production of the Msheireb Downtown CGIs demonstrates the global divisions of creative labor in a major urban redevelopment project (McNeill, 2008; see also Ren, 2011). At its very early stages, for example, the master planners (a London-based company) were tasked with developing a code for a specific architectural and visual language for the entire development. Much of the imaginative, creative work was led from studios in London and Liverpool on the basis of extensive research into the Qatari vernacular undertaken by the British consultants with a Qatari colleague, which was reviewed by a panel of US-based academics of mainly Middle Eastern origin. From that research, the master planners Allies and Morrison in London refined and developed the idea of a "contemporary Qatari" architectural idiom to inform a design code that responded to the character mission evoked by the Qatar Foundation and discussed in Chapter 3. This code was published as one of five volumes of Sitewide Design Guidelines. It governed the work of all the architects subsequently appointed to the project, imposing, we were told, a broad collective identity on the team selected through the architectural competition to design individual buildings and the master plan.

In this competition, twelve practices drawn from both Europe and the Middle East/Gulf region were invited to submit ideas, which ranged from the pastiche and orientalizing (including some of the regional entries) to the technocratic and over-scaled. The winning proposals were selected on the basis of their ability to understand and reinterpret the intimate quality of the traditional *fereej* or neighborhood typology found in Qatari and Arab towns and cities, within the overarching framework of a modern European or American-style gridded master plan with underground servicing. All but one of the winning practices were London-based (the exception was headquartered in Barcelona). This meant that all the designers involved were tasked with conjuring visions of a place in a city that the vast majority had never visited.

The production of the forty-two CGIs discussed in Chapter 3 was embedded in this distributed network of aesthetic labor in London (and much of the visualization work was created by a visualization studio in Liverpool). These working spaces were often light and airy spaces, full of design paraphernalia: collections of books, displays of previous projects, ongoing project boards. Both architects and visualizers worked largely in silence, seated in long rows in front of their screens in large studios, usually wearing headphones and immersed in their own work. Drawings and images danced about on screen in response to each touch of the mouse. Staff worked long hours, striving to keep track of the frequent updates and annotations circulated through Google docs, or on marked-up PDFs, resulting in that enormous proliferation of digital files on office servers. This was a consequence of the meticulous selection and control demanded by the new urban aesthetic of glamour: "the creator must edit out discordant details that could break the spell—blemishes on then skin, spots on the windows, electrical wires crossing the façade" (Postrel 2003 quoted in Thrift, 2008: 15). Tight deadlines had to be met, at which point the finished files would be uploaded to shared servers and printed out at different scales for the project directors to take out to Doha for the scheduled meetings with the client and project managers.

These design meetings were often tense, dynamic, face-to-face sessions, where the images took center stage in the evolving discussion about the design, development, and presentation of the project, and dictated client decisions on sign-off at each work stage. The aesthetic labor that produced the CGIs in time for these meetings was often frantic and stressful. The ALA talked about the making of an aesthetic CGI as an art:

> And maybe it's not the last [version of a CGI] but the one before the last one when you have all the materials, all the light sort of agreed but it still looks very plastic, it does not have this atmosphere. But then there is this two or three days when the artists are actually doing the arts of putting some sort of, we'll call it the God's Eye, that's the sort of beams of lights coming through the window. You know, where they will put some people, work on the textures, work on the people that the image comes to life. And it's this, it's where they are not just purely rendering, putting the light, but when they make an art out of the image. Cos it is art, you know, it is painting in a way but just painting with computers and painting in Photoshop. But this is how you can distinguish sort of normal renderer from really great renderer. That they have this ability to make sort of almost flat or plastic looking image into something really special.

Making a CGI is about making "something really special then" then. But it is also working "for two or three days on the people, textures and light" to make the image "come to life."

This aesthetic labor can make us look at Figure 3.3, for example, rather differently. As Chapter 3 summarized, a CGI like the one reproduced as Figure 3.3 has gone through multiple stages of production in which all sorts of elements are inserted, removed, and edited: visualizers receive plans and massing studies created in Microstation or AutoCAD by the architects for individual buildings, and import them into the visualization software 3DS Max. Then a digital 3D model of the building is created, made up of a number of layers. Then, light, materials, and texture are added via the software's layer manager—the image still does not look realistic at this stage. The visualizer then "renders" the 3D model, translating it into a 2D image, that is to be converted from a working picture to something that looks more like a photographic "picture": "[r]endering, then, consists of allocating 'texture' (a colour, a density, a nature or function) to computerised objects" (Houdart, 2008: 52). The next step is for the visualizer to import these renders into Photoshop, and to add more detail: subtle light effects, clouds, people, and trees. This can be done either by "painting in" effects using the software or by pasting in photographic images from elsewhere (sourced either from online digital image banks or from the visualizer's own collection).

From this discussion, we can see that the experience of making these CGIs was very different from the aesthetics they were intended to convey. All this labor is elided by the aesthetics of glow and luxurious glamour. The digital imagery pictures an effortless lifestyle and is itself meant to look and feel effortless. Audiences are meant to be seduced by the glamour, caught up in the flow, and thrilled by the drama. The aesthetics of the images themselves work on us to make feeling anything else rather difficult. Nonetheless, we are suggesting that these images need to be experienced differently: not only as seductive feelings of flow, glamour, and drama but also as things that are labored over stage by stage, often in different places by different people. And that can be done by seeing them seamfully, as patched together by different people in different places. The aesthetic of the Msheireb Downtown CGIs was glamorous, for sure. But—as is always the case—that glamour was enabled by huge amounts of technical and aesthetic labor which was not glamorous at all. The intimate, small-scale, and leisurely images of Msheireb Downtown were created by a global, fast-paced, and sometimes quite brutal workflow. Chapter 5 made it clear that in some ways we are all aesthetic laborers. Aesthetic labor is not relegated to professionals such as architects, visualizers, renderers, or designers. As our discussion highlighted, the affordances of digital platforms such as

Instagram have made it possible for everyone to become an expert in visual choreography and manipulation.

Similarly, the apparently seamlessness of the new urban aesthetic of flow discussed in Milton Keynes can be disrupted by focusing on the labor that attempted—but also often failed—to create that flow. As Chapter 4 noted that many of the MK apps never got beyond the prototype stage; only two managed to "scale up" beyond MK. Apps and platforms often glitch (Leszczynski, 2020). And in the process of designing the apps, there may be moments for the anticipated "users" to become somewhat unruly, just as Rancière suggests. A good example of this in MK is the Age Friendly Map app. Some of the aesthetic labor that produced the app prototype was undertaken by an app developer, as noted in Chapter 4 and again in the previous section, and much of his work normalized particular forms of rational agency and mobile embodiment. However, others also worked on the app design: the Age Concern worker, for example, and also the group of potential users she brought to two design focus groups. The project manager at DVC recalled:

> as we found out in the focus group some of them found that quite, how do I put it, so they didn't want to rest, they didn't want to be told when they wanted to rest and they found it quite judgemental and we're saying, "Oh public access toilets they need" and they're like, "What are you saying, that we need to go to the toilet all the time?"

That is, the focus group participants resisted the implications being designed into the app that they were less technologically able, poor, easily tired, and less continent than other bodies. For many older people, mobility is strongly aligned with a sense of their own independence (Schwanen et al., 2012); in this case, claiming mobility for the older body was to challenge the implicit ageism of the app's data offer. What they wanted, it turned out, was much the same as many other users of the city center: information on parking spaces, for example. As Chapter 5 argued, digital technologies often invite an expanded sense of who labors to create the new urban aesthetic, and marking all those interventions in the digital interface is an important critical tactic.

Seamfulness, then, is a way to multiply the corporeal sensibilities entrained in the new urban aesthetic. It specifically focuses on aesthetic labor both professional and everyday, in studios and in streets and workshops. We suggest that, in the context of urban branding, making the labor of producing new urban aesthetics apparent is another way of suggesting that their aesthetic qualities occlude other ways of distributing the sensible. If the images are seen as made, then they can be remade, made differently, to evoke other kinds of urban living (Rose et al., 2014).

6 Conclusion

This chapter is an effort to both describe specific distributions of the sensible and then to re-describe them. Hence, we started each section by describing and defining storytelling, animation, and seamfulness to show how they produce particular experiences of the new urban aesthetic and then re-described them to make them show how they could offer a different account. By doing so we demonstrate the power relations inherent in these aesthetic practices which shape spatiotemporal and sensory relations in and of places. Together, they offer a critical vocabulary as a tool to *re*distribute the sensible.

However, it is important to point out that we should not restrict ourselves to these three critical terms; and future studies of different places, different mediations, and different bodies will very likely reveal more and diverse constellations. We are also not alone in these attempts. For example, Leszczynski (2020) has elaborated on the notion of the "glitch." Looking for glitches, she says, "train[s] the eye precisely on those space-times where platforms appear in ways that, on the surface, seem not quite right: unexpectedly, otherwise than anticipated, or failing to appear at all" (2020: 201). "Training the eye" suggests a particular kind of attentiveness that can be cultivated by critical scholarship, just like our terms of storytelling, animation, and seamfulness. Chung (2018) too, in her book on digital-visual effects in the global movie production industry, works up a methodology that literally looks differently at those movies by questioning the seamfulness that many digital visualizations portray and instead reveals the diverse processes and labor that go into these aesthetic constructs. What the three concepts we are putting forward suggest is that the new urban aesthetic is not only all-encompassing, fixed, or permanent but, as our brief concluding chapter will argue, also fluid, unpredictable, and differentiated.

7

Conclusion

The Differentiation and Potentialities of the New Urban Aesthetic

1 Introduction

Sensory urban experiences are felt in in many and manifold ways through bodily capacities, sensations, and imaginaries, and this is a process saturated with a productive kind of power. "What you come into contact with is shaped by what you do: bodies are orientated when they are occupied in time and space. Bodies are shaped by this contact with objects. What gets near is both shaped by what bodies do, and in turn affects what bodies can do" (Ahmed, 2007: 152). Bodies are central to what happens in cities, and they are the matter through which sensible forms of power crystallize. Swanton (2010) elaborates this point by illustrating how social difference—in his study, race—takes form immanently through the coming together of particular material and immaterial urban elements. Focusing on a road in the north of the UK, and examining the interactions between Asian taxi drivers and white customers there, he shows how race and racism come momentarily (and sensorially) into matter through the encounter of bodies and things. Swanton highlights the ways in which race and the "sensing and judgement of racial difference" is enrolled in different intensities, at different times, and through different constellations of cars, discourses, and bodies, among many other things, showing that "differentiation is at least as much about relations between bodies, things and spaces as it is about discourse" (Swanton, 2010: 461). He clearly emphasizes the embodied situatedness of urban experiences, highlighting the entanglements between and power relations produced by bodies, in particular places, at specific times. Power is deeply experienced through sensorial registers of visibility, audibility, tactility, and so on.

This book has been an effort to bring "relations between bodies, things and spaces" to bear on the digital mediation of contemporary cities. It is widely agreed that cities are the key sites for the reproduction of global capitalism, and that urban branding is central to this transformative dynamic (Arantes, 2019; Cronin and Hetherington, 2008; Harvey, 1989; Hubbard and Hall, 1998; Tzanelli, 2017; van Heur, 2010). In the twenty-first century, many cities are also being profoundly reconfigured—in diverse ways—by the deployment of many kinds of digital technologies (Barns, 2020). Indeed, "digital phenomena have radically transformed almost every aspect of human life. From economies to cultures to politics, there is almost no area that remains untouched by digital techniques, logics or devices" (Ash et al., 2018: 1). Both of these shifts entrain bodily experience. What we have added to current discussions about urban change and digital technologies, then, is an attention to the relations between bodies, things, and spaces. Specifically we have explored one particular aspect of these relations in the convergence of the urban and the digital: the centrality of sensory experiencing to the reconfiguration of how cities are being branded, designed, planned, built, and lived. Digital data, software, and devices increasingly mediate how city spaces are sensed and corporeally perceived, thus reconfiguring urban experience into specific forms of aesthetic sensibilities. This digitally mediated reconfiguration of what cities feel like is *the new urban aesthetic*.

We have developed the concept of the new urban aesthetic to capture a relation between three elements: sensory experience, digital technologies, and urban materialities. This relation is complex, and we have identified a range of ways in which it is differentiated. It emerges in different ways in different cities. It is labored over using different digital and other tools. It is experienced in different registers, which we describe as conceived, perceived, and lived. And it is organized through distinctive distributions of sensibilities and sensations. Because of this, the new urban aesthetic is not a fixed construction but a dynamic constellation, contingent on how bodies, technologies, and urban materialities interact. It is therefore always multiple, transforming, and evolving. The new urban aesthetics is not a uniform or hegemonic force but, instead, a theoretical tool to conceptualize how new aesthetic configurations express themselves in different locations and through particular powerful dynamics.

2 The New Urban Aesthetic: Reprise

To illustrate and examine some of the differentiated expressions of the new urban aesthetic, we explored three case studies of urban restructuring projects which strongly feature different kinds of digital mediations. Each case study revealed a different formation of the new urban aesthetic, which we described as glamour, flow, and drama. Glamour refers to the ways in which the aesthetic labor, to create CGIs, produces alluring sensations by enchanting audiences through an appeal to their sensory perceptions and imagination, while also signifying something out of reach, a "mysterious world that inspires a mood" (Postrel, 2013: 213). Flow on the other hand refers to how smart city apps generate sensations of constant corporeal movement, of fluid emergence, which assume a body "conceived of in terms of independence of movement and bodily functions; a body without physical and mental impairments" (Imrie, 2000: 1643). Drama refers to another manifestation of a new urban aesthetic, namely how place-related posts on Instagram—posted by both place branding organizations and diverse users—elevate ordinary moments from their mundanity into dramatized events through careful editing, formatting, and manipulation, since in the Insta-era, "[e]verything is perceived as a photo opportunity, and this constant state of mind produces new forms of experiencing everyday life" (Serafinelli, 2018: 8). As we have emphasized, the particular configurations which we have identified in Qatar, Milton Keynes, and London are not directly transferable to other places but each is a specific manifestation of the new urban aesthetic. Elaborating such different manifestations was the first tactic in our efforts to problematize the new urban aesthetic.

The second was to work with Böhme's (2003) discussion of the production of aesthetic value for profit, especially in regard to the design and commercialization of urban space, as a phenomena of late capitalism. In particular Böhme explores the importance of staging sensory atmospheres and the conscious arrangement of sensuous displays in the city, emphasizing the "aesthetic labor" implicated in these processes and of particular relevance to urban branding as we have illustrated. So, our second move was to be attentive to the work done by corporeal bodies to mediate aesthetically urban spaces: that is to different kinds of aesthetic labor. We analyzed the work done to create digital visualizations of a planned development (Chapter 3) and to design smartphone apps (Chapter 4), and looked at how a place branding company and Instagram users stage striking Instagram posts (Chapter 5). All of these examples reveal different forms of aesthetic production that involve different bodies, different urban social practices, different urban imaginaries, and different aesthetics. Böhme's work can be regarded as a continuation of the "critical theory" espoused by the Frankfurt School (Albertsen, 2016). Yet, while Böhme warns that the aesthetic economy is driven by a desire for the enhancement and "intensification" of life into a process of "exploitation" that captures people in dependency on limitless "escalation of desires" (Böhme, 2001: 183–4), he also argues that understanding how atmospheres are produced will "sharpen the critical potential and hence also the resilience (*Widerstandskraft*) against economic and also political manipulation" (Böhme, 2001: 52). Paying attention to the various kinds of aesthetic labor involved in each of our case studies showed just how much work is needed to produce new urban aesthetics—and also that the results of that work might not cohere or grip.

Our third tactic for problematizing the new urban aesthetic was to understand aestheticized urban environments as textured (Jansson, 2013). To do so we drew on Lefebvre's (1991) understanding of space as constituted through three distinct sociospatial dimensions constructed through bodies engaged in the building, making, representing, framing, and experiencing of space (Simonsen, 2005). This allowed us to highlight the different forms and relations between how the new urban aesthetic is *conceived* in multiple ways by different urban actors; *perceived* in manifold ways through diverse bodies, urban materialities, and technologies; and *lived* through manifold practices, both everyday and specialized. Again, these different dimensions may not align.

The fourth tactic was to work with Rancière's (2004) conception of the "distribution of the sensible" to link an analysis of spatial power to sensory experiencing. Rancière delineates how the way we perceive the world is based on a range of socially regulated sensory frameworks. He further argues that this organization of the sensible depends on the configurations of spaces and times which shape "the place and the stake of politics as a form of experience" (Rancière, 2004: 13). We examined how glamorous, flowing, and dramatic configurations of urban space and time make only certain sensations sensible, or tend to respond to particular kinds of bodily sensibility, or only intensify specific bodily dispositions. Each of our empirical chapters demonstrates different versions of this process. Drawing on Rancière's notion of the distribution of the sensible allowed us to elaborate a distinctive approach to the power dynamics embedded in the new urban aesthetic, namely how aesthetics are organized by particular configurations of urban spatiality and temporality which constitute different embodied experiences in particular neighborhoods and cities.

In Chapter 6 we discussed the different distributions of the sensible as a form of power relations. We developed a vocabulary to interrogate the new urban aesthetic which identified both how it was policed and how the work done to achieve that policing might be incomplete. The three terms we worked with were storytelling, animation, and seamfulness. We are sure that future studies will reveal others. We discussed how these features both produce and simultaneously displace possible formations of the sensible in space and time and, thereby, we paid attention to what is rendered marginal or not possible and to what is excluded and excessive to a new urban aesthetic. Describing power through the workings of sensory aesthetics in relation to digital mediations of the urban opens up new possibilities to think through the manifold ways in which power can be interrupted, challenged, and remade (Jordan and Lindner, 2016). Distributions of the sensible are uneven, contradictory, and often fragile. Our emphasis on the feel of cities, their sensory experiencing, and atmospheres "offers an orientation to encounter that is never fully defined or completed and therefore remains open to unknown turns and sudden developments" (Sumartojo and Pink, 2019: 120). Important for the argument of this book is that the fluidity and dynamism of the experiential world provide opportunities for multiplicity, unpredictability, frictions, excesses, glitches, surpluses, or resistances that underpin the configuration of the new urban aesthetic. To conclude the book, we would like to offer just a few more examples that underline some of these unpredictable possibilities.

3 The Limits of the New Urban Aesthetic

In Smithfield, we spoke to a teenage hairdresser in one of the local businesses of the area earmarked as the Culture Mile. She told us how she found the area gray and monotonous. Asked whether she could recommend any cafés locally, she told us her boss had ordered her not to leave the shop and thus she was not aware of any cafés. This stood in stark contrast with tourists, hip visitors, and creative industry workers we interviewed in the area. The sensorial framings produced through the digital mediations of Instagram of the Culture Mile were also not part of this woman's experience. While we have emphasized how Instagram posts shape how cities are conceived, perceived, and lived in different ways for different social positions (see also Boy and Uitermark, 2017, 2020), it is important not to forget that the reach of such posting is limited. Hence, while urban aesthetic power works through corporeality, Böhme (2007: 26) explains that we are not helplessly exposed to these atmospheric powers: we can glance casually and let them go (Krajina, 2014), or simply be unaware of them, as the example of the hairdresser shows. This points to the multiplicity of possible urban experiences of course. It also emphasizes the unevenness of the new urban aesthetic. It does not entrain all bodies.

The failure of a new urban aesthetic to grip can be evident even in the process of its production. Creating the Msheireb Downtown CGIs, for example, became fraught when their seamfulness started to disrupt their material production. In other work we have used the notion of friction because it points to the failures that are inherent in digital technologies: friction materializes and disrupts human-digital relations. As we noted at the beginning of this book, much of the digital mediation of urban experience occurs via screen interfaces: the smartphone, the computer, and the app. But friction is inherent in interfaces since the interface is "an autonomous zone of interaction . . . concerned as much with unworkability and obfuscation as with connectivity and transparency" (Galloway, 2012: 120). Thus, errors in the system or in the way humans interact with technology tend to happen and digital things are "vulnerable to the frictions of crashes, glitches and error" (Rose, 2016: 364). During the production of the Msheireb Downtown images, many difficulties and challenges emerged that had to be resolved by aesthetic labor(ers) (Rose et al., 2014). Examples range from office computers that were not able to open the huge image files of complex digital renders; confusion over what version of a visualization was to be worked on; and instructions on how a visualization had to be altered not being understood when received. Different kinds of friction affect the different components of the interface and network.

In the case of Msheireb Downtown, these frictions were generally resolved. In her analysis of digital platform-urban configurations, Leszczynski (2020) develops the notion of glitch for frictions that persist (see also Elwood, 2020). She argues that as well as its reference to failure or malfunction, such as when there is an error in code or misunderstandings between humans, glitches have an inherent propensity to critical potential. So, for example, we have outlined in Chapter 3 how the production of digital visualizations such as those created to picture the Msheireb Downtown project are based on a highly complex iterative process that involves a range of actors. As the CGIs were repeatedly sent to and fro between the architects, visualizers, and client representatives, a "comment system" was used. Specific instructions for visualizers were written on a CGI printout by the ALA with a red pen, or typed as comments on a PDF file: "more light," "more people," "bicycle out," "magic please"—mainly referring to the feelings and sensory atmospheres conjured in the images. This process might happen several times, and through each modification and added or altered layer of texture, color, light, and detail, objective criteria and consensus about decisions become more elusive. Indeed, the whole process was fraught with tensions due to the inherent ambivalence of aesthetic atmospheres and the subjective interpretations they demand (Anderson, 2009). Thus, we saw a house visualizer looking baffled as he looked at an annotated PDF and saying: "I don't understand. What does he (ALA) mean 'more magic'? He wants like a man flipping cards? I don't know!." Our ability to witness such deliberations by paying attention to the labor of creating these visualizations suggests that they are far from being near-magical, seamless, pristine image of glossy urban futures: "instead, they are rather more like sites of debate and disagreement which shift and change as different designs are inputted, different sorts of views desired, and different sorts of audiences anticipated" (Rose et al., 2016). As a result, glitches occur.

4 Other New Aesthetics of Cities

The new urban aesthetic then is differentiated. It materializes differently in different places, it is textured, it must be labored over and it distributes sensibilities in distinctive configurations. Its textures, labor, and distributions have limits, unevennesses, glitches, frictions, instabilities, and excesses. This offers not only opportunities for disruption but also "creative openings . . . that allow for a reconceptualization of what can (or cannot) be realised within existing [socio-digital] practices" (Nunes, 2011: 4).

As Chapter 5 indicated, the efforts of professional aesthetic labor employed to brand particular places in particular ways are also part of a much wider field of digital labor (Hearn and Banet-Weiser, 2020). This labor is part of urban lived experience for very many people now. In the wider context of platform urbanism, many of us live in intimate entanglements with very many kinds of digital technologies (Barns, 2020). While it is the case that much of the digital labor we have been attentive to in this book is deeply complicit with both aesthetic capitalism and platform capitalism, we also want to gesture here toward the ways in which these complicities can be challenged. Many artists, activists, and campaigners work to make the materialities of urban digital transformations evident, whether that involves mapping urban digital infrastructure (Mattern, 2016) or highlighting the seamfulness of digital image. Artist and digital activist James Bridle has drawn attention to the constructed and photoshopped nature of CGIs by creating "render ghosts," for example (Bridle, 2013). He creates close-ups of the human figures photoshopped into CGIs, which make visible traces of the digital cut-and-paste operation that inserted them into the visualization, thus making them look "out of place" and disrupting the dazzling and immersive spectacle that CGIs try to evoke. Photographers track the decay of billboards like the one in Figure 3.3, as wind and dust and traffic fumes do their own re-rendering of the image into something dirty and tattered; others collate the absurdity of those billboards' marketing slogans (Rose et al., 2016).

And talking of the absurd, we also want to go beyond these artist-led interventions and flag the fecund creativity of vast arena of more everyday digital labor that goes on now. Funny, silly, ridiculous, idiotic (Goriunova, 2013): from memes to Vines to Tik-Toks, there is a vast amount of satire, comedy, play, and pleasure in the everyday digital mediation of cities that is precisely unpredictable (Hartley, 2012). It hosts the unanticipated. This is not to ignore the ways that this demotic creativity can also be vicious, exclusionary, and abusive. But the digital is also a site for creative expression, including the articulation of identities otherwise marginalized, as Brock (2020) demonstrates in his discussion of "distributed blackness" (and see

Datta and Thomas, 2021). A small example of the unpredictability inherent in this everyday aesthetic labor can be seen in an unexpected moment of the Culture Mile branding via Instagram. As one of the marketing officers told us, the Culture Mile organized a weekend event to celebrate 150 years of Smithfield Market. During this street-party, all kinds of activities took place, from music performances to fun fairs and henna drawing. Although the event took place during a warm August Bank Holiday weekend and was well attended, it started to rain just when the "sausage dog" competition started. As the marketing director explained while the public got soaked, no one seemed to remember the downpour but, rather surprisingly for the Culture Mile organizers, one of the most posted images over that weekend was the sausage dog competition. Complicity with the eventfulness of the dramatic new urban aesthetic? Pleasure in everyday oddness? Or random silliness? Whatever, its unpredictability directly disrupts the Culture Mile project's efforts to predict the experiencing of Smithfield; its unanticipatedness is a glitch in efforts to anticipate that area's urban future.

Like many others, then, this book has argued that digital mediation is crucial to the development and experiencing of urban environments in contemporary cities. Digital technologies mediate "how we experience being with both human and non-human digital others in the spaces and practices of everyday life, in which we come to understand spaces, experiences, and interactions as the effects of the myriad comings-together of technology, society, and space relations" (Leszczynski, 2018: 18). If we are living in an era of platform capitalism, then platform urbanism is also a thing. However, the current era has also been described as an era of aesthetic capitalism. What we add to urban scholarship on digitally mediated cities is a focus on the sensory implications of the intersection of the digital and the aesthetic in urban spaces. Especially in the context of the imperative to brand urban places as redeveloped or smart, digital mediations take powerful and intensely *aesthetic* forms. While there are clear intersections between aesthetic capitalism and platform capitalism in terms of the extraction and commodification of data (Hearn and Banet-Weiser, 2020), urban scholarship has not yet paid much attention to the specific dynamics of aesthetic power in the digitally mediated city. By revealing how this new urban aesthetic can be conceptualized as differentiated in many ways, we hope to have shifted the conversation toward a more nuanced analysis that seeks to understand the diverse experiencing of cities. The new urban aesthetic is situated, differential, and multiple. Its power lies in its policing of urban sensations. And as with all policing, the spatial and temporal organization of its sensory distributions require much labor to hold, and so can fail, be side-stepped, breached, ignored, and displaced.

5 Afterword

The latter part of writing for this book took place during the Covid-19 pandemic. The pandemic and its repercussions on urban life, the economy, tourism, and all of our individual lives are still to be determined. In relation to our broader arguments about urban spaces, digital mediations, and sensory experiences, we have certainly witnessed the start of new phenomena where people perform music or sport from their balconies (Hassan, 2020; Locker and Hoffman, 2020), or use their front windows to support local health services (Greig, 2020), or craft festive advent windows (Osborne, 2020)—and share all these on social media. Moreover, different kinds of digital tracking devices have been deployed in many cities globally to trace bodies in close contact with each other, and social networks have both expanded and contracted when reliant on videoconferencing technologies. This only highlights how novel figurations of the new urban aesthetic, the ways in which we experience urban change digitally, will continue to emerge. What the pandemic has thrown into light is how quickly urban experiences can be reframed.

References

Abella A, Ortiz-de-Urbina-Criado M and De Pablos-Heredero C (2015) The reuse of information in smart cities' ecosystems. *El Profesional de la Informacion* 24(6): 6838–44.

Abram S and Wezkalnys G (2011) Anthropologies of planning – Temporality, imagination and ethnography. *Focaal – Journal of Global and Historical Anthropology* 61: 3–18.

Acuti D, Mazzoli V, Donvito R, et al. (2018) An instagram content analysis for city branding in London and Florence. *Journal of Global Fashion Marketing* 9(3): 185–204.

Adam B (1990) *Time and Social Theory*. Oxford: John Wiley & Sons.

Adam B (2008) Of timescapes, futurescapes and timeprints. Conference paper delivered at Lüneburg University, 2008.

Adams M, Moore G, Cox T, et al. (2007) The 24-hour city: Residents' sensorial experiences. *The Senses and Society* 2(2): 201–15.

Ahmed S (2007) A phenomenology of whiteness. *Feminist Theory* 8(2): 149–68.

Albertsen N (2016) Atmosphere: Power, critique, politics. A conceptual analysis. In: *Proceedings of 3rd International Congress on Ambiances*, Volos, Greece, September 2016, pp. 573–78.

Allen J (2006) Ambient power: Berlin's potsdamer platz and the seductive logic of public spaces. *Urban Studies* 43(2): 441–55.

Amin A (2014) Lively infrastructure. *Theory, Culture & Society* 31(7–8): 137–61.

Amoore L and Poitukh V (eds) (2016) *Algorithmic Life*. Abingdon: Routledge.

Andéhn M, Kazeminia A, Lucarelli A, et al. (2014) User-generated place brand equity on Twitter: The dynamics of brand associations in social media. *Place Branding and Public Diplomacy* 10(2): 132–44.

Anderson B (2009) Affective atmospheres. *Emotion, Space and Society* 2(2): 77–81.

Anderson B (2010) Preemption, precaution, preparedness: anticipatory action and future geographies. *Progress in Human Geography* 34(6): 777–98.

Anderson D (2017) *Imaginary Cities*. Chicago: University of Chicago Press.

Arantes P (2019) *The Rent of Form: Architecture and Labour in the Digital Age*. Minneapolis: University of Minnesota Press.

Ash J (2010) Architectures of affect: Anticipating and manipulating the event in processes of videogame design and testing. *Environment and Planning D: Society and Space* 28(4): 653–71.

Ash J (2015) Technology and affect: Towards a theory of inorganically organised objects. *Emotion, Space and Society* 14(2): 84–90.

Ash J, Kitchin R and Leszczynski A (eds) (2018) *Digital Geographies*. London: Sage.

Ashworth GJ and Voogd H (1990) *Selling the City: Marketing Approaches in Public Sector Urban Planning*. London: Belhaven Press.

Aurigi A and Willis K (eds) (2020) *The Routledge Companion to Smart Cities*. Abingdon: Routledge.

Balibrea MP (2017) *The Global Cultural Capital*. Basingstoke: Palgrave MacMillan.

Barns S (2018) We are all platform urbanists now. *Mediapolis* 3(No 4 Roundtables). Available at: https://www.mediapolisjournal.com/2018/10/we-are-all-platform-urbanists-now/.

Barns S (2020) *Platform Urbanism: Negotiating Platform Ecosystems in Connected Cities*. Basingstoke: Palgrave Macmillan.

Bassano C, Barile S, Piciocchi P, et al. (2019) Storytelling about places: Tourism marketing in the digital age. *Cities* 87: 10–20.

Benjamin R (2019) *Race After Technology: Abolitionist Tools for the New Jim Code*. Cambridge: Polity Press.

Berry C, Harbord J and Moore R (eds) (2013) *Public Space, Media Space*. Basingstoke: Palgrave Macmillan.

Bissell D (2010) Passenger mobilities: Affective atmospheres and the sociality of public transport. *Environment and Planning D: Society and Space* 28(2): 270–89.

Bissell D and Fuller G (2017) Material politics of images: Visualising future transport infrastructures. *Environment and Planning A* 49(11): 2477–96.

Böhme G (1993) Atmosphere as the fundamental concept of a new aesthetics. *Thesis Eleven* 36(1): 113–26.

Böhme G (2001) *Aisthetik. Vorlesungen Über Ästhetik Als Allgemeine Wahrnehmungslehre*. München; Paderborn: Wilhelm Fink Verlag.

Böhme G (2003) Contribution to the critique of the aesthetic economy. *Thesis Eleven* 73(1): 71–82.

Böhme G (2007) Technical gadgetry. Technological development in the aesthetic economy. In: Heil R, Kaminski A, Stippak M, et al. (eds) *Tensions and Convergences: Technological and Aesthetic Transformations of Society*. Bielefeld: Transcript, pp. 23–35.

Böhme G (2013) The art of the stage set as a paradigm for an aesthetics of atmospheres. *Ambiances: Environnement Sensible, Architecture et Espace Urbain*: 1–8. Available at: https://journals.openedition.org/ambiances/315

Böhme G (2014) Urban atmospheres: Charting new directions for architecture and urban planning. In: Borch C (ed.) *Architectural Atmospheres: On the Experience and Politics of Architecture*. Basel: Walter de Gruyter GmbH, pp. 42–59.

Böhme G (2017) *Atmospheric Architectures: The Aesthetics of Felt Spaces* (tran. A-C Engels-Schwarzpaul). London: Bloomsbury.

Bonakdar A and Audirac I (2019) City branding and the link to urban planning: Theories, practices, and challenges. *Journal of Planning Literature* 35(2): 147–60.

BOP Consulting and Publica (2019) *Culture Mile Annual Report*. BOP Consulting & Publica.

Boy JD and Uitermark J (2017) Reassembling the city through Instagram. *Transactions of the Institute of British Geographers* 42(4): 612–24.

Boy JD and Uitermark J (2020) Lifestyle enclaves in the Instagram city? *Social Media + Society* 6(3): 1–10.

Boyer M-C (1988) The return of aesthetics to city planning. *Society* 25: 49–56.

Boyer M-C (1996) *CyberCities: Visual Perception in the Age of Electronic Communication*. New York: Princeton Architectural Press.
Brenner N and Theodore N (eds) (2003) *Spaces of Neoliberalism: Urban Restructuring in North America and Western Europe*. Oxford: Wiley-Blackwell.
Brenner N, Peck J and Theodore N (2010) After neoliberalization? *Globalizations* 7(3): 327–45.
Bridle J (2013) The render ghosts. In: *Electronic Voice Phenomena*. Available at: http://www.electronicvoicephenomena.net/index.php/the-render-ghosts-james-bridle/ (accessed January 2, 2021).
Brighenti AM and Pavoni A (2020) Vertical vision and atmocultural navigation. Notes on emerging urban scopic regimes. *Visual Studies* 35(5): 429–41.
Brock A (2020) *Distributed Blackness: African American Cybercultures*. New York: New York University Press.
Brown J (2009) *Glamour in Six Dimensions: Modernism and the Radiance of Form*. New York: Cornell University Press.
Browne S (2015) *Dark Matters: On the Surveillance of Blackness*. London: Duke University Press.
Buckingham S, Degen M and Marandet E (2018) "Lived bodies" and the neoliberal city – a case study of vulnerability in London. *Gender, Place & Culture* 25(3): 334–350.
Buser M (2017) Atmospheres of stillness in Bristol's Bearpit. *Environment and Planning D: Society and Space* 35(1): 126–45.
Butler A, Schafran A and Carpenter G (2018) What does it mean when people call a place a shithole? Understanding a discourse of denigration in the United Kingdom and the Republic of Ireland. *Transactions of the Institute of British Geographers* 43(3): 496–510.
Caliandro A and Graham J (2020) Studying Instagram beyond selfies. *Social Media + Society* 6(2): 1–7.
Calzada I (2018) (Smart) citizens from data providers to decision-makers? The case study of Barcelona. *Sustainability* 10(9): 3252.
Carah N and Shaul M (2016) Brands and Instagram: Point, tap, swipe, glance. *Mobile Media & Communication* 4(1): 69–84.
Cardullo P and Kitchin R (2018) Smart urbanism and smart citizenship: The neoliberal logic of "citizen-focused" smart cities in Europe. The Programmable City Working Paper Series. Available at: https://osf.io/preprints/socarxiv/xugb5.
Cardullo P and Kitchin R (2019) Smart urbanism and smart citizenship: The neoliberal logic of "citizen-focused" smart cities in Europe. *Environment and Planning C: Politics and Space* 37(5): 813–30.
Casetti F (2015) *The Lumière Galaxy: Seven Key Words for the Cinema to Come*. New York: Columbia University Press.
Castells M (1996) *The Rise of The Network Society*. Oxford: Blackwell.
Chalmers M (2003) Seamful design and ubicomp infrastructure. *Proceedings Ubicomp 2003 Workshop At the Crossroads: The Interaction of HCI and Systems Issues in UbiComp*. Available at: http://www.dcs.gla.ac.uk/~matthew/papers/ubicomp2003HCISystems.pdf.
Cheney-Lippold J (2017) *We Are Data: Algorithms and the Making of Our Digital Selves*. New York: New York University Press.
Chien S and Woodworth MD (2018) China's urban speed machine: The politics of speed and time in a period of rapid urban growth. *International Journal of Urban and Regional Research* 42(4): 723–737.

Christophers B (2016) For real: land as capital and commodity. *Transactions of the Institute of British Geographers* 41(2): 134–48.
Chung HJ (2018) *Media Heterotopias: Digital Effects and Material Labor in Global Production*. London: Duke University Press.
City of London Corporation (2018) *City of London Cultural Strategy 2018–2022*. London: City of London Corporation.
Clough PT (2018) *The User Unconscious: Affect, Media and Measure*. London: University of Minnesota Press.
Cockayne DG (2016) Entrepreneurial affect: Attachment to work practice in San Francisco's digital media sector. *Environment and Planning D: Society and Space* 34(3): 456–73.
Cowley R and Caprotti F (2019) Smart city as anti-planning in the UK. *Environment and Planning D: Society and Space* 37(3): 428–48.
Coyne R (2010) *The Tuning of Place: Sociable Spaces and Pervasive Digital Media*. Cambridge: MIT Press.
Crang M (2001) Temporalised space and motion. In: May J and Thrift N (eds) *Timespace: Geographies of Temporality*. London: Routledge, pp. 187–207.
Crang M, Crosbie T and Graham S (2006) Variable geometries of connection: Urban digital divides and the uses of information technology. *Urban Studies* 43(13): 2551–70.
Crang M, Crosbie T and Graham S (2007) Technology, time–space, and the remediation of neighbourhood life. *Environment and Planning A: Economy and Space* 39(10): 2405–22.
Cronin A and Hetherington K (eds) (2008) *Consuming the Entrepreneurial City: Image, Memory, Spectacle*. Abingdon: Routledge.
Cultural Hub Working Party Report to Policy and Resources Committee. London: City of London Corporation.
Culture Mile (2020) *Brand Guidelines*. London: Culture Mile.
Culture Mile London website (2020). Available at: https://www.culturemile.london/.
Datta A (2016) Introduction: Fast cities in an urban age. In: Datta A and Shaban A (eds) *Mega-Urbanization in the Global South: Fast Cities and New Urban Utopias of the Postcolonial State*. Abingdon: Routledge.
Datta A (2018) The digital turn in postcolonial urbanism: Smart citizenship in the making of India's 100 smart cities. *Transactions of the Institute of British Geographers* 43(3): 405–19.
Datta A (2020) The "Smart Safe City": Gendered time, speed, and violence in the margins of India's urban age. *Annals of the American Association of Geographers* 110(5): 1318–34.
Datta A and Shaban A (2016) *Mega-Urbanization in the Global South: Fast Cities and New Urban Utopias of the Postcolonial State*. Abingdon: Routledge.
Datta A and Thomas A (2021) Curating #AanaJaana [#ComingGoing]: Gendered authorship in the "contact zone" of Delhi's digital and urban margins. *Cultural Geographies* online first.
de Souza e Silva A and Frith J (2012) *Mobile Interfaces in Digital Spaces: Locational Privacy, Control, and Urban Sociability*. Abingdon: Routledge.
De Waal M (2013) *The City as Interface: How New Media Are Changing the City*. Rotterdam: NAI Uitgevers/Publishers Stichting.
Dean J (2010) *Blog Theory: Feedback and Capture in the Circuits of Drive*. Cambridge: Polity Press.

Degen M (2008) *Sensing Cities: Regenerating Public Life in Barcelona and Manchester*. London: Routledge.
Degen M (2010) Consuming urban rhythms: Let's ravalejar. In: Edensor T (ed.) *Geographies of Rhythm*. Aldershot: Ashgate.
Degen M (2014) The everyday city of the senses. In: *Cities and Social Change: Encounters with Contemporary Urbanism*. London: Sage, pp. 92–111.
Degen M (2017) Urban regeneration and "resistance of place": Foregrounding time and experience. *Space and Culture* 20(2): 141–55.
Degen M (2018) Timescapes of urban change: The temporalities of regenerated streets. *The Sociological Review* 66(5): 1074–92.
Degen M and Barz M (2019) Mapping urban experience digitally. In: Ward K (ed.) *Researching the City: A Guide for Students*. Abingdon: Routledge.
Degen M and Lewis C (2020) The changing feel of place: The temporal modalities of atmospheres in Smithfield Market, London. *Cultural Geographies* 27(4): 509–26.
Degen M and Rose G (2012) The sensory experiencing of urban design: The role of walking and perceptual memory. *Urban Studies* 49(15): 3269–85.
Degen M, Melhuish C and Rose G (2017) Producing place atmospheres digitally: Architecture, digital visualisation practices and the experience economy. *Journal of Consumer Culture* 17(1): 3–24.
Degen M, DeSilvey C and Rose G (2008) Experiencing visualities in designed urban environments: Learning from Milton Keynes. *Environment and Planning A: Economy and Space* 40(8): 1901–20.
Dekeyser T (2018) The material geographies of advertising: Concrete objects, affective affordance and urban space. *Environment and Planning A: Economy and Space* 50(7): 1425–42.
Deleuze G (1992) Postscript on the societies of control. *October* 59: 3–7.
Dieter M, Gerlitz C, Helmond A, et al. (2019) Multi-situated app studies: Methods and propositions. *Social Media + Society* 5(2): 1–15.
Dikec M (2015) *Space, Politics and Aesthetics*. Edinburgh: Edinburgh University Press.
Dorrian M (2008) 'The way the world sees London': Thoughts on millennial urban spectacle. In: Vidler A (ed.) *Architecture between Spectacle and Use*. London: Yale University Press.
Ebbensgaard C (2015) Illuminights: A sensory study of illuminated urban environments in Copenhagen. *Space and Culture* 18(2): 112–31.
Ebbensgaard C (2020) Standardised difference: Challenging uniform lighting through standards and regulation. *Urban Studies* 57(9): 1957–76.
Edensor T (2012) Illuminated atmospheres: Anticipating and reproducing the flow of affective experience in Blackpool. *Environment and Planning D-Society & Space* 30(6): 1103–22.
Elsaesser T (2013) The "return" of 3-D: On some of the logics and genealogies of the image in the twenty-first century. *Critical Inquiry* 39(2): 217–46.
Elwood S (2020) Digital geographies, feminist relationality, Black and queer code studies: Thriving otherwise. *Progress in Human Geography* 45(2): 209–28.
Ernwein M and Matthey L (2018) Events in the affective city: Affect, attention and alignment in two ordinary urban events. *Environment and Planning A: Economy and Space* 51(2): 283–301.
Eshuis J and Edwards A (2013) Branding the city: The democratic legitimacy of a new mode of governance. *Urban Studies* 50(5): 1066–82.

Eubanks V (2017) *Automating Inequality: How High-Tech Tools Profile, Police, and Punish the Poor*. New York: St Martin's Press.

Farman J (2015) Infrastructures of mobile social media. *Social Media + Society* 1(1): 1–12.

Fast K, Jansson A, Tesfahuney M, et al. (2018) Introduction to geomedia studies. In: Fast K, Jansson A, Lindell J, et al. (eds) *Geomedia Studies: Spaces and Mobilities in Mediatized Worlds*. Basingstoke: Routledge, pp. 1–18.

Fessler L (2018) So long, sexism: Amazon's Alexa is now a feminist, and she's sorry if that upsets you. *Quartz*, January 17. Available at: https://qz.com/work/1180607/amazons-alexa-is-now-a-feminist-and-shes-sorry-if-that-upsets-you/.

Fields D, Bissell D and Macrorie R (2020) Platform methods: Studying platform urbanism outside the black box. *Urban Geography* 41(3): 462–8.

Finn E (2017) *What Algorithms Want: Imagination in the Age of Computing*. London: MIT Press.

Florida R (2005) *Cities and the Creative Class*. New York: Routledge.

Frers L (2016) Perception, aesthetics, and envelopment–encountering space and materiality. In: Frers L and Meier L (eds) *Encountering Urban Places: Visual and Material Performances in the City*. Aldershot: Ashgate, pp. 41–62.

Friedman AT (2010) *American Glamour and the Evolution of Modern Architecture*. London: Yale University Press.

Frosh P (2003) *The Image Factory: Consumer Culture, Photography and the Visual Content Industry*. London: Berg.

Gabrys J (2014) Programming environments: Environmentality and citizen sensing in the smart city. *Environment and Planning D: Society and Space* 32(1): 30–48.

Gabrys J, Pritchard H and Barratt B (2016) Just good enough data: Figuring data citizenships through air pollution sensing and data stories. *Big Data & Society* 3(2): 205395171667967.

Galloway A (2012) *The Interface Effect*. Cambridge: Polity Press.

Gandy M (2017) Urban atmospheres. *Cultural Geographies* 24(3): 353–74.

Georgiou M (2013) *Media and the City: Cosmopolitanism and Difference*. Cambridge: Polity Press.

Ghertner A (2015) *Rule by Aesthetics: World-Class City Making in Delhi*. New York: Oxford University Press.

Gieseking JJ (2017) Messing with the attractiveness algorithm: A response to queering code/space. *Gender, Place & Culture* 24(11): 1659–65.

Gordon E (2010) *The Urban Spectator: American Concept Cities from Kodak to Google*. Hanover: Dartmouth College Press.

Goriunova O (2013) New media idiocy. *Convergence: The International Journal of Research into New Media Technologies* 19(2): 223–35.

Graham M (2020) Regulate, replicate, and resist – The conjunctural geographies of platform urbanism. *Urban Geography*: 41(3): 453–7.

Graham M and Anwar MA (2019) The global gig economy: Towards a planetary labour market? *First Monday* 24(4): 1–31.

Graham M, Zook M and Boulton A (2013) Augmented reality in urban places: Contested content and the duplicity of code. *Transactions of the Institute of British Geographers* 38(3): 464–79.

Graham S and Marvin S (2001) *Splintering Urbanism: Networked Infrastructures, Technological Mobilities and the Urban Condition*. London: Routledge.

Greenberg M (2000) A social history of the urban lifestyle magazine. *Urban Affairs Review* 36(2): 228–63.
Greenberg M (2007) *Branding New York: How a City in Crisis Was Sold to the World*. New York: Routledge.
Greenfield A (2013) *Against the Smart City*. London: Do Projects.
Greig F (2020) Rainbows in windows. *I News*, April 3.
Grossi G and Pianezzi D (2017) Smart cities: Utopia or neoliberal ideology? *Cities* 69: 79–85.
Hakala U, Lätti S and Sandberg B (2011) Operationalising brand heritage and cultural heritage. *Journal of Product and Brand Management* 20(6): 447–56.
Halegoua GR (2019) *The Digital City: Media and The Social Production of Place*. New York: New York University Press.
Halpern O (2015) *Beautiful Data: A History of Vision and Reason since 1945*. Durham: Duke University Press.
Hartley J (2012) *Digital Futures for Cultural and Media Studies*. Chichester: John Wiley.
Harvey D (1989a) From managerialism to entrepreneurialism: The transformation in urban governance in late capitalism. *Geografiska Annaler: Series B, Human Geography* 71(1): 3–17.
Harvey D (1989b) *The Condition of Postmodernity*. Oxford: Blackwell.
Harvey D (1990) Between space and time: Reflections on the geographical imagination. *Annals of the Association of American Geographers* 80(3): 418–34.
Harvey D (2005) *Spaces of Neoliberalization: Towards a Theory of Uneven Geographical Development*. Stuttgart: Franz Steiner Verlag.
Hassan M (2020) During quarantine, balconies worldwide set the stage for DJ sets, squats and singing. *The Washington Post*, March 16. Available at: https://www.washingtonpost.com/world/2020/03/16/under-quarantine-balconies-around-world-set-stage-dj-sets-squats-singing/ (accessed December 17, 2020).
Hayles NK (2004) Print is flat, code is deep: The importance of media-specific analysis. *Poetics Today* 25(1): 67–90.
Hearn A and Banet-Weiser S (2020) The beguiling: Glamour in/as platformed cultural production. *Social Media + Society* 6(1): 1–11.
Herzog LA (1995) Fast urbanism and slow urbanism: Globalization and public space in three Mexican cities. In: *The End of Public Space in the Latin American City*, Austin: University of Texas.
Hetherington K (2013) Rhythm and noise: The city, memory and the archive. *The Sociological Review* 61(S1): 17–33.
Hildebrand JM and Sheller M (2018) Media ecologies of autonomous mobility. *Transfers* 8(1): 64–85.
Ho E (2017) Smart subjects for a Smart Nation? Governing (smart)mentalities in Singapore. *Urban Studies* 54(13): 3101–18.
Hoch C (2009) Planning craft: How planners compose plans. *Planning Theory* 8(3): 219–41.
Hoelzl I and Marie R (2015) *Soft Image: Towards a New Theory of the Digital Image*. Bristol: Intellect.
Hollands RG (2008) Will the real smart city please stand up? *City* 12(3): 303–20.
Houdart S (2008) Copying, cutting and pasting social spheres: Computer designers' participation in architectural projects. *Science Studies: An Interdisciplinary Journal for Science and Technology Studies* 21(1): 47–63.

Howes D (2005) *Empire of the Senses*. Oxford: Berg Publishers.
Huawei (2017) Bristol-overtakes-London-as-UK-smartest-city. Available at: https:/ /www.huawei.com/uk/news/uk/2017/huawei-uk-smart-cities-index (accessed August 28, 2020).
Hubbard P and Hall T (1998) The entrepreneurial city and the 'new urban politics'. In: *The Entrepreneurial City: Geographies of Politics, Regime and Representation*. London: John Wiley & Sons, pp. 1–23.
Hubbard P and Lyon D (2018) Introduction: Streetlife–the shifting sociologies of the street. *The Sociological Review* 66(5): 937–51.
Huopalainen A (2019) Manipulating surface and producing "effortless" elegance – analysing the social organization of glamour. *Culture and Organization* 25(5): 332–52.
Imrie R (2000) Disability and discourses of mobility and movement. *Environment and Planning A: Economy and Space* 32(9): 1641–56.
Isin EF and Ruppert E (2015) *Being Digital Citizens*. London: Rowman & Littlefield.
Iveson K and Maalsen S (2018) Social control in the networked city: Datafied dividuals, disciplined individuals and powers of assembly. *Environment and Planning D: Society and Space* 37(2): 331–49.
Jansson A (2013) Mediatization and social space: Reconstructing mediatization for the transmedia age. *Communication Theory* 23(3): 279–96.
Jefferson BJ (2017) Digitize and punish: Computerized crime mapping and racialized carceral power in Chicago. *Environment and Planning D: Society and Space* 35(5): 775–96.
Jenkins H, Ford S and Green J (2013) *Spreadable Media: Creating Value and Meaning in a Networked Culture*. New York: New York University Press.
Jensen OB (2007) Culture stories: Understanding cultural urban branding. *Planning Theory* 6(3): 211–36. DOI: 10.1177/1473095207082032.
John N (2016) *The Age of Sharing*. Cambridge: Polity Press.
Jordan S and Lindner C (eds) (2016) *Cities Interrupted: Visual Culture and Urban Space*. London: Bloomsbury.
Joss S, Cook M and Dayot Y (2017) Smart cities: Towards a new citizenship regime? A discourse analysis of the British Smart City Standard. *Journal of Urban Technology* 24(4): 29–49.
Karaliotas L and Swyngedouw E (2019) Exploring insurgent urban mobilizations: From urban social movements to urban political movements? In: Schwanen T and van Kempen R (eds) *Handbook of Urban Geography*. Cheltenham: Edward Elgar, pp. 369–82.
Karvonen A and van Heur B (2014) Urban laboratories: Experiments in reworking cities. *International Journal of Urban and Regional Research* 38(2): 379–92.
Kavaratzis M (2009) Cities and their brands: Lessons from corporate branding. *Place Branding and Public Diplomacy* 5(1): 26–37.
Kember S and Zylinska J (2012) *Life After New Media: Mediation as a Vital Process*. Cambridge: MIT Press.
Kenyon N (2018) The Culture Mile that will transform arts in the City. *Evening Standard*, March 16. Available at: https://www.standard.co.uk/comment/comment/the-culture-mile-that-will-transform-arts-in-the-city-a3792026.html (accessed January 2, 2020).
Kincaid E (2020) Re-encountering Lefebvre: Toward a critical phenomenology of social space. *Environment and Planning D: Society and Space* 38(1): 167–86.

Kinsley S (2010) Representing 'things to come': Feeling the visions of future technologies. *Environment and Planning A* 42(11): 2771–90.
Kinsley S (2012) Futures in the making: Practices to anticipate 'ubiquitous computing'. *Environment and Planning A* 44(7): 1554–69.
Kitchin R (2014) The real-time city? Big data and smart urbanism. *GeoJournal* 79(1): 1–14.
Kitchin R (2017) Thinking critically about and researching algorithms. *Information, Communication & Society* 20(1): 14–29.
Kitchin R and Dodge M (2011) *Code/Space: Software and Everyday Life*. Cambridge MA: MIT Press.
Kitchin R, Maalsen S and McArdle G (2016) The praxis and politics of building urban dashboards. *Geoforum* 77: 93–101.
Kitchin R, Lauriault TP and McArdle G (eds) (2017) *Data and the City*. Abingdon: Routledge.
Kitchin R, Claudio C, Leighton E, et al. (2017) Smart cities, epistemic communities, advocacy coalitions and the "last mile" problem. *it - Information Technology* 59(6): 275–84.
Klingmann A (2007) *Brandscapes: Architecture in the Experience Economy*. London: MIT Press.
Krajina Z (2014) *Negotiating the Mediated City: Everyday Encounters with Public Screens*. Abingdon: Routledge.
Krivý M (2016) Towards a critique of cybernetic urbanism: The smart city and the society of control. *Planning Theory* 17(1): 8–30.
Kuntsman A (2012) Introduction. In: Karatzogianni A and Kuntsman A (eds) *Digital Cultures and The Politics of Emotion: Feelings, Affect and Technological Change*. Basingstoke: Palgrave Macmillan, pp. 1–20.
Langley P and Leyshon A (2017) Platform capitalism: The intermediation and capitalization of digital economic circulation. *Finance and Society* 3(1): 11–31.
Lash S and Lury C (2007) *Global Culture Industry*. Cambridge: Polity Press.
Latham A and McCormack DP (2004) Moving cities: Rethinking the materialities of urban geographies. *Progress in Human Geography* 28(6): 701–24.
Law R and Underwood K (2012) Msheireb heart of Doha: An alternative approach to urbanism in the Gulf Region. *International Journal of Islamic Architecture* 1(1): 131–47.
Leaver T, Highfield T and Abidin C (2020) *Instagram: Visual Social Media Cultures*. Cambridge: Polity Press.
Lee A, Mackenzie A, Smith GJD, et al. (2020) Mapping platform urbanism: Charting the nuance of the platform pivot. *Urban Planning* 5(1): 116–28.
Lefebvre H (1991) *The Production of Space*. Oxford: Blackwell.
León LFA and Rosen J (2020) Technology as ideology in urban governance. *Annals of the American Association of Geographers* 110(2): 497–506.
Leszczynski A (2016) Speculative futures: Cities, data, and governance beyond smart urbanism. *Environment and Planning A* 48(9): 1691–708.
Leszczynski A (2018) Spatialities. In: Ash J, Kitchin R, and Leszczynski A (eds) *Digital Geographies*. London: Sage, pp. 13–23.
Leszczynski A (2019) Platform affects of geolocation. *Geoforum* 107: 207–15.
Leszczynski A (2020) Glitchy vignettes of platform urbanism. *Environment and Planning D: Society and Space* 38(2): 189–208.
Levitt D (2018) *The Animatic Apparatus: Animation, Vitality, and the Futures of the Image*. Winchester: Zero Books.

Lew AA (2017) Tourism planning and place making: Place-making or placemaking? *Tourism Geographies* 19(3) Routledge: 448–66.
Lindner C and Sandoval GF (eds) (2021) *Aesthetics of Gentrification: Seductive Spaces and Exclusive Communities in the Neoliberal City*. Amsterdam: Amsterdam University Press.
Lindstrom M (2005) *Brand Sense: Sensory Secrets Behind the Stuff We Buy*. New York: The Free Press.
Locker M and Hoffman A (2020) People quarantined in Italy join together in song from balconies during Coronavirus. *Time*, March 15. Available at: https://time.com/5802700/lockdown-song/ (accessed March 15, 2020).
Lorentzen A (2009) Cities in the experience economy. *European Planning Studies* 17(6): 829–45.
Lucarelli A (2018) Place branding as urban policy: The (im)political place branding. *Cities* 80: 12–21.
Lund NF, Cohen SA and Scarles C (2018) The power of social media storytelling in destination branding. *Journal of Destination Marketing & Management* 8: 271–80.
Lupton D (2016) *The Quantified Self*. London: John Wiley.
Lupton D (2018) How do data come to matter? Living and becoming with personal data. *Big Data & Society* 5(2): 2053951718786314.
Lupton D (2019) *Data Selves: More-than-Human Perspectives*. London: John Wiley & Sons.
Luque-Ayala A and Marvin S (2020) *Urban Operating Systems: Producing the Computational City*. London: MIT Press.
Luque-Ayala A and Neves Maia F (2018) Digital territories: Google maps as a political technique in the re-making of urban informality. *Environment and Planning D: Society and Space* 37(3): 449–67.
MacKenzie A and Munster A (2019) Platform seeing: Image ensembles and their invisualities. *Theory, Culture & Society* 36(5): 3–22.
Mackrodt U (2019) How atmospheres inform urban planning practice – insights from the Tempelhof airfield in Berlin. *Ambiances. Environnement Sensible, Architecture et espace urbain* 5. Available at: https://journals.openedition.org/ambiances/2739.
Manovich L (2016) What is digital cinema? In: Denson S and Leyda J (eds) *Post-Cinema: Theorizing 21st-Century Film*. Falmer: Reframe, pp. 20–50.
Manovich L (2017) *Instagram and Contemporary Image*. Manovich.net. Available at: http://manovich.net/index.php/projects/instagram-and-contemporary-image.
Markusen A and Gadwa A (2010) *Creative Placemaking*. Washington, DC: National Endowment for the Arts.
Mattern S (2015) Mission control: A history of the urban dashboard. Available at: https://placesjournal.org/article/mission-control-a-history-of-the-urban-dashboard/ (accessed March 13, 2015).
Mattern S (2016) Cloud and field: On the resurgence of "field guides" in a networked age. *Places Journal* (August). Available at: https://placesjournal.org/article/cloud-and-field/ (accessed September 7, 2016).
Mattern S (2016) Indexing the world of tomorrow. *Places Journal*. Available at: https://placesjournal.org/article/indexing-the-world-of-tomorrow-1939-worlds-fair/.
Mattern S (2017) *Code and Clay, Data and Dirt: Five Thousand Years of Urban Media*. Minneapolis: University of Minnesota Press.

McLuhan M (1964) *Understanding Media: The Extensions of Man*. London: Routledge and Kegan Paul.
McNeill D (2008) *The Global Architect: Firms, Fame and Urban Form*. London: Routledge.
McNeill D (2015) Global firms and smart technologies: IBM and the reduction of cities. *Transactions of the Institute of British Geographers* 40(4): 562–74.
McNeill D (2016) IBM and the visual formation of smart cities. In: Marvin S, Luque-Ayala A, and McFarlane C (eds) *Smart Urbanism: Utopian Vision or False Dawn?* London and New York: Routledge.
McQuire S (2008) *The Media City: Media, Architecture and Urban Space*. London: Sage.
McQuire S (2010) Rethinking media events: Large screens, public space broadcasting and beyond. *New Media & Society* 12(4): 567–82.
McQuire S (2016) *Geomedia: Networked Cities and the Future of Public Space*. Cambridge: Polity Press.
Melhuish C, Degen M and Rose G (2014) Architectural atmospheres: Affect and agency of mobile digital images in the material transformation of the urban landscape in Doha. *Tasmeem* 4. Available at: https://www.qscience.com/content/journals/10.5339/tasmeem.2014.4.
Melhuish C, Degen M and Rose G (2016) "The real modernity that is here": Understanding the role of digital visualisations in the production of a new urban imaginary at Msheireb Downtown, Doha. *City & Society* 28(2): 222–45.
Moores S (2014) Digital orientations: "Ways of the hand" and practical knowing in media uses and other manual activities. *Mobile Media & Communication* 2(2): 196–208.
Mould O (2015) *Urban Subversion and the Creative City*. Abingdon: Routledge.
Msheireb Properties (2020) Msheireb Downtown Doha - Mandate. Available at: https://www.msheireb.com/msheireb-downtown-doha/about-msheireb-downtown-doha/mandate/ (accessed November 6, 2020).
Munster A (2006) *Materializing New Media: Embodiment in Information Aesthetics*. Hanover, NH: Dartmouth College Press, University Press of New England.
Munster A (2013) *An Aesthesia of Networks: Conjunctive Experience in Art and Technology*. London: MIT Press.
Museum of London Strategic Plan 2018–2023 (2018) Museum of London. Available at: museumoflondon.org.uk/strategicplan.
Myers D and Kitsuse A (2000) Constructing the future in planning: A survey of theories and tools. *Journal of Planning Education and Research* 19: 221–31.
Ngai S (2010) Our aesthetic categories. *PMLA/Publications of the Modern Language Association of America* 125(4): 948–58.
Nicodemus AG (2013) Fuzzy vibrancy: Creative placemaking as ascendant US cultural policy. *Cultural Trends* 22(3–4): 213–22.
Noble SU (2018) *Algorithms of Oppression: How Search Engines Reinforce Racism*. New York: New York University Press.
November V, Camacho-Hübner E and Latour B (2010) Entering a risky territory: space in the age of digital navigation. *Environment and Planning D: Society and Space* 28(4): 581–99.
Nunes M (2011) *Error: Glitch, Noise, and Jam in New Media Cultures*. London: Continuum.

Osborne H (2020) 'It brings cheer after a rubbish year': Advent windows light up UK streets. *Guardian*, December 12. Available at: https://www.theguardian.com/lifeandstyle/2020/dec/12/advent-windows-light-up-uk-streets (accessed December 17, 2020).

Otter C (2008) *The Victorian Eye: A Political History of Light and Vision in Britain, 1800–1910*. London: University of Chicago Press.

Paasonen S (2016) Fickle focus: Distraction, affect and the production of value in social media. *First Monday* 21(10): 1–11.

Paklone I (2011) Conceptualization of visual representation in urban planning. *LIMES: Cultural Regionalistics* 4(2): 150–61.

Papastergiadis N, McQuire S, Gu X, et al. (2013) Mega screens for mega cities. *Theory, Culture & Society* 30(7–8): 325–41.

Pickford J (2017) Nicholas Kenyon on London's new 'Culture Mile'. *Financial Times*, July 20.

Pine BJ and Gilmore JH (1999) *The Experience Economy: Work Is Theatre and Every Business a Stage*. Boston: Harvard Business School.

Pollio A (2016) Technologies of austerity urbanism: The "smart city" agenda in Italy (2011–2013). *Urban Geography* 37(4): 514–34.

Postrel V (2013) *The Power of Glamour: Longing and the Art of Visual Persuasion*. New York: Simon and Schuster.

Prakash G (2008) Introduction. In: Prakash G and Kruse KM (eds) *The Spaces of the Modern City: Imaginaries, Politics, and Everyday Life*. Oxford: Princeton University Press, pp. 1–18.

Publica (2015) *Vision for the City of London's Cultural Hub*. Publica and City of London Corporation.

Raco M (2014) Delivering flagship projects in an era of regulatory capitalism: State led privatization and the London Olympics 2012. *International Journal of Urban and Regional Research* 38(1): 176–97.

Raco M, Henderson S and Bowlby S (2008) Changing times, changing places: Urban development and the politics of space–time. *Environment and Planning A: Economy and Space* 40(11): 2652–73.

Rancière J (2004) *The Flesh of Words: The Politics of Writing* (tran. C Mandell). Stanford: Stanford University Press.

Rancière J (2006) *The Politics of Aesthetics: The Distribution of the Sensible* (tran. G Rockhill). London: Continuum.

Rancière J (2009) *The Emancipated Spectator* (tran. G Elliott). London: Verso.

Redstrom J and Wiltse H (2018) *Changing Things: The Future of Objects in a Digital World*. London: Bloomsbury.

Ren X (2011) *Building Globalization: Transnational Architecture Production in Urban China*. Chicago: University of Chicago Press.

Richards G (2014) Creativity and tourism in the city. *Current Issues in Tourism* 17(2) Routledge: 119–44.

Richardson L (2020a) Platforms, markets, and contingent calculation: The flexible arrangement of the delivered meal. *Antipode* 52(3): 619–36.

Richardson L (2020b) Coordinating the city: Platforms as flexible spatial arrangements. *Urban Geography* 41(3): 458–61.

Roberts M and Eldridge A (2012) *Planning the Night-Time City*. Abingdon: Routledge.

Rodgers S and Moore S (2018) Platform urbanism: An introduction. *Mediapolis* 3(4). Available at: https://www.mediapolisjournal.com/2018/10/platform-urbanism-an-introduction/ (accessed March 6, 2020).

Rodgers S, Barnett C and Cochrane A (2014) Media practices and urban politics: Conceptualizing the powers of the media–urban nexus. *Environment and Planning D: Society and Space* 32(6): 1054–70.

Rose G, Degen M and Melhuish C (2016) Looking at digital visualizations of urban redevelopment projects: Dimming the scintillating glow of unwork. In: Jordan S and Lindner C (eds) *Cities Interrupted: Visual Culture and Urban Space*. London: Bloomsbury, pp. 105–20.

Rose G (2016) Rethinking the geographies of cultural "objects" through digital technologies: Interface, network and friction. *Progress in Human Geography* 40(3): 334–51.

Rose G (2017a) Posthuman agency in the digitally mediated city: Exteriorization, individuation, reinvention. *Annals of the American Association of Geographers* 107(4): 779–93.

Rose G (2017b) Screening smart cities: Managing data, views and vertigo. In: Hesselberth P and Poulaki M (eds) *Compact Cinematics: The Moving Image in the Age of Bit-Sized Media*. London: Bloomsbury Academic, pp. 177–84.

Rose G (2018) Look InsideTM: Visualising the smart city. In: Fast K, Jansson A, Lindell J, et al. (eds) *An Introduction to Geomedia: Spaces and Mobilities in Mediatized Worlds*. Abingdon: Routledge, pp. 97–113.

Rose G (2019a) Pinterest: Curating a city. In: Graham M, Kitchin R, Mattern S, et al. (eds) *How To Run a City Like Amazon and Other Fables*. Oxford: Meatspace Press, pp. 676–705.

Rose G (2019b) Smart urban: Intelligence, interiority, imagination. In: Lindner C and Meissner M (eds) *The Routledge Companion to Urban Imaginaries*. Abingdon: Routledge, pp. 105–12.

Rose G (2021) Representational and animatic corporeality: Refiguring bodies and digitally mediated cities. In: Million A, Haid C, Castillo Ulloa I, et al. (eds) *Spatial Transformations: The Effect of Mediatization, Mobility, and Social Dislocation on the Re-Figuration of Spaces*. Basingstoke: Routledge.

Rose G, Degen M and Melhuish C (2014) Networks, interfaces, and computer-generated images: Learning from digital visualisations of urban redevelopment projects. *Environment and Planning D: Society and Space* 32(3): 386–403.

Rose G and Willis A (2019) Seeing the smart city on Twitter: Colour and the affective territories of becoming smart. *Environment and Planning D: Society and Space* 37(3): 411–27.

Rose G, Degen M and Basdas B (2010) More on 'big things': Building events and feelings. *Transactions of the Institute of British Geographers* 35(3): 334–49.

Rose G, Degen M and Melhuish C (2014) Networks, interfaces, and computer-generated images: Learning from digital visualisations of urban redevelopment projects. *Environment and Planning D: Society and Space* 32(3): 386–403.

Rose G, Degen M and Melhuish C (2015) Looking at digital visualisations of urban redevelopment projects: Dimming the scintillating glow of unwork. In: Jordan S and Lindner C (eds) *Cities Interrupted: Visual Culture, Globalisation and Urban Space*. London: Bloomsbury, pp. 105–20.

Rose G, Raghuram P, Watson S and Wigley E (2021) Platform urbanism, smartphone applications and valuing data in a smart city. *Transactions of the Institute of British Geographers*, 46(1): 59–72.

Rotenberg V (2012) Towards a genealogy of downtown. In: McDonagh G and Peterson M (eds) *Global Downtowns*. Philadelphia: University of Pennsylvania Press.

Rouvroy A (2013) The end(s) of critique: Data behaviourism versus due process. In: Hildebrandt M and Vries K de (eds) *Privacy, Due Process and the Computational Turn: The Philosophy of Law Meets the Philosophy of Technology*. Abingdon: Routledge.
Rubinstein D and Sluis K (2008) A life more photographic: Mapping the networked image. *Photographies* 1(1): 9–28.
Sadowski J (2020) Cyberspace and cityscapes: On the emergence of platform urbanism. *Urban Geography*: 1–5. DOI: 10.1080/02723638.2020.1721055.
Sadowski J and Bendor R (2018) Selling smartness: Corporate narratives and the smart city as a sociotechnical imaginary. *Science, Technology, & Human Values* 44(3): 540–63.
Sadowski J and Pasquale F (2015) The spectrum of control: A social theory of the smart city. *First Monday* 20(7). Available at: http://firstmonday.org/ojs/index.php/fm/article/view/5903 (accessed September 6, 2018).
Saisselin RG (1984) *The Bourgeois and the Bibelot*. New Brunswick: Rutgers University Press.
Sassen S (1991) *The Global City: New York, London, Tokyo*. Princeton: Princeton University Press.
Sassen S (2012) Interactions of the technical and the social. *Information, Communication & Society* 15(4): 455–78.
Sassen S (2016) The global city: Enabling economic intermediation and bearing its costs. *City & Community* 15(2): 97–108.
Schindler S and Marvin S (2018) Constructing a universal logic of urban control? *City* 22(2): 298–307.
Schwanen T (2015) Beyond instrument: Smartphone app and sustainable mobility. *European Journal of Transport and Infrastructure Research* 15(4): 675–90.
Schwanen T, Banister D and Bowling A (2012) Independence and mobility in later life. *Geoforum* 43(6): 1313–22.
Serafinelli E (2018) *Digital Life on Instagram: New Social Communication of Photography*. Bingley: Emerald Publishing.
Sevin E (2013) Places going viral: Twitter usage patterns in destination marketing and place branding. *Journal of Place Management and Development* 6(3): 227–39.
Sharma S (2014) *In the Meantime: Temporality and Cultural Politics*. Durham NC: Duke University Press.
Shelton T (2017) The urban geographical imagination in the age of Big Data. *Big Data & Society* 4(1): 2053951716665129.
Shelton T and Lodato T (2019) Actually existing smart citizens: Expertise and (non) participation in the making of the smart city. *City* 23(1): 35–52.
Simonsen K (2005) Bodies, sensations, space and time: The contribution from Henri Lefebvre. *Geografiska Annaler: Series B, Human Geography* 87(1): 1–14.
Simonsen K (2010) Encountering O/other bodies: Practice, emotion and ethics. In: Anderson B and Harrison P (eds) *Taking-Place: Non-Representational Theories and Geography*. Farnham: Ashgate, pp. 221–39.
Simonsen K (2013) In quest of a new humanism: Embodiment, experience and phenomenology as critical geography. *Progress in Human Geography* 37(1): 10–26.
Simonsen K and Koefoed L (2020) *Geographies of Embodiment*. London: Sage.
Simpson P (2016) A sense of the cycling environment: Felt experiences of infrastructure and atmospheres. *Environment and Planning A: Economy and Space* 49(2): 426–47.

Smart Cities UK (2020) City & sector rankings – Smart cities UK 2020. Available at: https://www.smartcityuk.com/resources/2020/1/13/city-amp-sector-rankings (accessed August 28, 2020).

Smith K (2019) 50 incredible Instagram statistics you need to know. Available at: https://www.brandwatch.com/blog/instagram-stats/ (accessed December 31, 2020).

Söderström O, Paasche T and Klauser F (2014) Smart cities as corporate storytelling. *City* 18(3): 307–20.

Soja EW (1996) *Thirdspace: Expanding the Geographical Imagination*. Oxford: Blackwell.

Soja EW (2000) *Postmetropolis: Critical Studies of Cities and Regions*. Oxford: Blackwell.

Srnicek N (2014) Auxiliary organs: An ethics of extended mind. In: Miller PD and Matviyenko S (eds) *The Imaginary App*. London: MIT Press, pp. 71–82.

Srnicek N (2016) *Platform Capitalism*. Cambridge: Polity Press.

Strengers Y and Kennedy J (2020) *The Smart Wife: Why Siri, Alexa, and Other Smart Home Devices Need a Feminist Reboot*. Cambridge: MIT Press.

Sumartojo S and Pink S (2019) *Atmospheres and The Experiential World: Theory and Methods*. Abingdon: Routledge.

Sumartojo S, Pink S, Lupton D, et al. (2016) The affective intensities of datafied space. *Emotion, Space and Society* 21: 33–40.

Sumartojo S, Edensor T and Pink S (2019) Atmospheres in urban light. *Ambiances. Environnement sensible, architecture et espace urbain* 5. Available at: https://journals.openedition.org/ambiances/2586.

Summers BT (2021) Race, authenticity, and the gentrified aesthetics of belonging in Washington, D.C. In: Lindner C and Sandoval GF (eds) *Aesthetics of Gentrification: Seductive Spaces and Exclusive Communities in the Neoliberal City*. Amsterdam: Amsterdam University Press, pp. 115–36.

Swanton D (2010) Sorting bodies: Race, affect, and everyday multiculture in a Mill Town in Northern England. *Environment and Planning A: Economy and Space* 42(10): 2332–50.

Swedberg R (2011) The role of the senses ad signs in the economy. *Journal of Cultural Economy* 4(4): 423–36.

Swyngedouw E, Moulaert F and Rodriguez A (2002) Neoliberal urbanization in Europe: Large–scale urban development projects and the new urban policy. *Antipode* 34(3): 542–77.

Taylor NB and While A (2017) Competitive urbanism and the limits to smart city innovation: The UK Future Cities initiative. *Urban Studies* 54(2): 501–19.

Thatcher J (2017) You are where you go, the commodification of daily life through 'location'. *Environment and Planning A: Economy and Space* 49(12): 2702–17.

Thatcher J, O'Sullivan D and Mahmoudi D (2016) Data colonialism through accumulation by dispossession: New metaphors for daily data. *Environment and Planning D: Society and Space* 34(6): 990–1006.

The Independent (2017) UK a nation of smartphone addicts with more than half saying they use theirs too much, research shows. Available at: https://www.independent.co.uk/news/business/news/uk-a-nation-of-smartphone-addicts-with-more-than-half-saying-they-use-theirs-too-much-research-shows-a7955701.html (accessed July 26, 2018).

Thelander Å and Cassinger C (2017) Brand new images? Implications of Instagram photography for place branding. *Media and Communication* 5(4): 6–14.

Thrift N (2008) The material practices of glamour. *Journal of Cultural Economy* 1(1): 9–23.

Thrift N (2012) The insubstantial pageant: Producing an untoward land. *Cultural Geographies* 19(2): 141–68.

Thrift N (2014) The "sentient" city and what it may portend. *Big Data & Society* 1(1): 1–21.

Thrift N and French S (2002) The automatic production of space. *Transactions of the Institute of British Geographers* 27(3): 309–35.

Thylstrup NB and Teilmann S (2017) Thumbnail images: Uncertainties, infrastructures and search engines. *Digital Creativity* 28(4): 279–96.

Till J (2009) *Architecture Depends*. Cambridge: MIT Press.

Tolia-Kelly DP (2006) Affect – an ethnocentric encounter? Exploring the "universalist" imperative of emotional/affectual geographies. *Area* 38(2): 213–17.

Toscano P (2017) Instagram-City: New media, and the social perception of public spaces. *Visual Anthropology* 30(3): 275–86.

Tzanelli R (2017) *Mega-Events as Economies of the Imagination: Creating Atmospheres for Rio 2016 and Tokyo 2020*. London: Routledge.

Valdez A-M, Cook M and Potter S (2018) Roadmaps to utopia: Tales of the smart city. *Urban Studies* 55(15): 3385–403.

van Heur B (2010) *Creative Networks and the City: Towards a Cultural Political Economy of Aesthetic Production*. transcript. Available at: https://doi.org/10.14361/transcript.9783839413746.

Vanolo A (2014) Smartmentality: The smart city as disciplinary strategy. *Urban Studies* 51(5): 883–98.

Verhoeff N (2012) *Mobile Screens: The Visual Regime of Navigation*. Amsterdam: Amsterdam University Press.

Wainwright O (2018) Snapping point: How the world's leading architects fell under the Instagram spell. *The Guardian*, 23 November.

Weber R (2010) Selling city futures: The financialization of urban redevelopment policy. *Economic Geography* 86(3): 251–74.

White JM (2016) Moving applications: A multilayered approach to mobile computing. In: Kitchin R and Perng S-Y (eds) *Code and the City*. Abingdon: Routledge, pp. 130–45.

Wigley E and Rose G (2020) Who's behind the wheel? Visioning the future users and urban contexts of connected and autonomous vehicle technologies. *Geografiska Annaler: Series B, Human Geography* (online first): 1–17.

Wiig A (2016) The empty rhetoric of the smart city: From digital inclusion to economic promotion in Philadelphia. *Urban Geography* 37(4): 535–53.

Willis K and Aurigi A (2018) *Digital and Smart Cities*. Abingdon: Routledge.

Wilmott C (2016) In-between mobile maps and media: Movement. *Television & New Media* 18(4): 320–35.

Wissinger E (2007) Always on display: Affective production in the modelling industry. In: Clough PT and Halley J (eds) *The Affective Turn: Theorising the Social*. Durham NC: Duke University Press, pp. 231–60.

Withers D (2015) *Feminism, Digital Culture and the Politics of Transmission*. London: Rowman & Littlefield International.

Woodman R (2012) City making: Doha. *Building Design Online*. Available at: https://www.bdonline.co.uk/city-making-doha/5034818.article (accessed July 5, 2020).

Yaneva A (2009) *The Making of a Building: A Pragmatist Approach to Architecture*. Oxford: Peter Lang.

Zappavigna M (2016) Social media photography: Construing subjectivity in Instagram images. *Visual Communication* 15(3): 271–92.

Zuboff S (2019) *The Age of Surveillance Capitalism: The Fight for A Human Future at the New Frontier of Power*. London: Profile.

Zuccala A (2018) New plans for the city's culture mile approved. *City Matters*, October 29. Available at: https://www.citymatters.london/plans-city-culture-mile-approved/ (accessed November 24, 2020).

Zukin S (2011) Reconstructing the authenticity of place. *Theory and Society* 40(2): 161–5.

Zukin S (2020) Seeing like a city: How tech became urban. *Theory and Society* 49(506): 941–64.

Index

www.ingramcontent.com/pod-product-compliance
Lightning Source LLC
LaVergne TN
LVHW010605110826
845149LV00003B/780

9781350283510